International Cooperation

A volume in the series

Cornell Studies in Political Economy

EDITED BY PETER J. KATZENSTEIN

A full list of titles in the series appears at the end of the book

International Cooperation

Building Regimes for Natural Resources and the Environment

Oran R. Young

Cornell University Press

Ithaca and London

First published 1989 by Cornell University Press.
First printing, Cornell Paperbacks, 1989.
Third printing 1993.

International Standard Book Number (cloth) 0-8014-2214-0
International Standard Book Number (paper) 0-8014-9521-0
Library of Congress Catalog Card Number 88-19078
Printed in the United States of America
Librarians: Library of Congress cataloging information
appears on the last page of the book.

⊗ The paper in this book meets the minimum requirements of the American National Standard for Information Sciences—Permanence of Paper for Printed Library Materials, ANSI Z39.48-1984.

Contents

CONTENTS

vi

Preface

The ideas this book presents regarding international regimes and, more generally, institutional arrangements in international society have been a long time in the making. I first seized on the term *regime* in 1973 while working on a study of the international relations of resource management with particular reference to the North Pacific or, to be more precise, the region known to natural scientists and archeologists as Beringia. This study resulted in a paper for the conference "Lawmaking in the Global Community" at Princeton University in 1975 and culminated in a book entitled *Resource Management at the International Level: The Case of the North Pacific* (Frances Pinter Ltd. and Nichols Publishing Co., 1977), in which I spoke of a regime "as a system of governing arrangements for a given social structure or region" (p. i). While I no longer use quite the same vocabulary in characterizing regimes, this early formulation did capture a good deal of the essence of what regime analysis has subsequently sought to illuminate.

Even so, my early usage of the term *regime* was rather casual. The concept was a handy device for organizing a mass of observations about working arrangements in a stateless social system (that is, international society). But it did not, at that stage, have much in the way of conceptual or analytic underpinnings. At the time, therefore, it did not occur to me that regime analysis would become a lively focus of attention among students of international relations.

The next phase in the evolution of my thinking about regimes occurred in a different context. In 1977, I began teaching courses on the political economy of natural resources. In this connection, I sought to pinpoint the determinants of social or group choices regarding the human use of natural resources and environmental services. The search led me increasingly to zero in on structures of property rights, lesser property interests (such as usufructuary rights, mineral leases, and limited-entry permits), and liability rules—in short, institutional arrangements. In the process, I came to see resource regimes as a subset of social institutions and to realize that this linkage would open up a rich vein of conceptual and theoretical ideas that could be brought to bear on the study of regimes. These efforts culminated in a book on the political economy of natural resources entitled *Resource Regimes: Natural Resources and Social Institutions* (University of California Press, 1982).

Meanwhile, my interest in the role of institutional arrangements in international society grew steadily. Above all, this line of thinking offered a means of interpreting the remarkable degree of stability or order that exists in international society, despite the absence of conventional governmental agencies at the international level. My efforts along these lines began to come together in an essay entitled "International Regimes: Problems of Concept Formation," which I prepared originally for presentation at the annual meetings of the International Studies Association in 1979 and which now forms the basis of the argument I present in chapter 1 of this book. But my thinking about international regimes really crystalized in the course of my participation in workshops on international regimes held in Los Angeles in October 1980 and in Palm Springs in February 1981. These workshops not only stimulated my thinking about institutional arrangements in international society, they also helped me to see how my own evolving ideas about international regimes differed from the ideas of a number of my colleagues. For this, I owe a debt of gratitude to Stephen Krasner, who was the principal organizer of these workshops and who edited the volume entitled *International Regimes* (Cornell University Press, 1983) which emerged from these sessions.

In the course of the 1980s, these separate streams of analysis converged to produce the present volume on international regimes for natural resources and the environment. I have no doubt that issues pertaining to natural resources and the environment are now on the international agenda to stay. The dramatic growth in public awareness of problems involving radioactive fallout, acid deposition, the depletion of stratospheric ozone, and the global warming trend has seen to this. Moreover, I am now convinced that regime analysis offers an appropriate vehicle for tackling problems of this kind. Applied in a rigorous and innovative fashion, these tools can help us to devise institutional arrangements to cope with transboundary environmental problems as well as to deepen our comprehension of the bases of stability or order in international society. Accordingly, I hope this book will serve as an intellectual springboard for those beginning to grapple with the rapidly expanding agenda of transboundary environmental issues. Certainly, I expect to follow this line of enquiry myself during the foreseeable future.

I take this opportunity to express my thanks to all those who have contributed to my thinking about international regimes over the years. Assorted colleagues have helped me to escape at least some of the pitfalls of this research program over the better part of two decades. I am especially grateful to those who have resisted the whole idea of regime analysis or persisted in pointing out vague and imprecise elements in this way of thinking. These colleagues are too numerous to mention individually here; I owe a large debt of gratitude to all those who regularly contribute their time and energy to ensure that an invisible college operates effectively among students of international relations.

I want particularly to offer thanks to those who helped in the final phase of transforming my work on international regimes into the present book. Each time we met in recent years, my longtime friend Nicholas Onuf provided impetus by invariably asking when I was going to get around to preparing a book of this sort. Roger Haydon of Cornell University Press triggered this final transformation by writing to enquire whether I was intending to produce a book on international regimes. Friedrich Kratochwil and Peter Katzenstein, who read the first draft of the completed manuscript for Cornell University Press, were enormously helpful in persuading me to make substantial revisions in preparing the final draft. While exhausted authors do not always welcome suggestions for serious revisions at this stage, I am especially grateful to them for the clarity and incisiveness of their criticisms. Though I alone am responsible for the remaining shortcomings of the book, I can say with certainty that the final product would have been far less satisfactory without the help I received from all these colleagues along the way.

ORAN R. YOUNG

Wolcott, Vermont

Acknowledgments

Chapter 1 is based on a paper presented at the annual meetings of the International Studies Association in March 1979; an earlier version appeared as "International Regimes: Problems of Concept Formation," *World Politics* 32, no. 3 (April 1980), 331–356. Copyright © 1980 by Princeton University Press. Adapted with permission of Princeton University Press. Chapter 2 is a revision of a paper presented at the American Political Science Association annual meetings in August 1986. Chapter 3 originated as a presentation at the annual meetings of the International Studies Association in April 1987. Chapter 4 began as a paper for a workshop on international regimes organized by Stephen Krasner; an earlier version appeared as "Regime Dynamics: The Rise and Fall of International Regimes," in *International Organization* 36, no. 2 (1982), 277–297, and in Stephen D. Krasner, ed., *International Regimes* (Ithaca: Cornell University Press, 1983), 93–113. Adapted with permission of The MIT Press and the World Peace Foundation.

Chapter 5 is a much-revised version of an essay prepared for a conference at Resources for the Future in Washington, D.C., and published as "International Resource Regimes," in Clifford S. Russell, ed., *Collective Decision Making: Applications from Public Choice Theory* (Baltimore: Johns Hopkins University Press for Resources for the Future, 1979), 241–282. Copyright © 1979 by Resources for the Future. The writing of an earlier version of chapter 6 was supported by The Annenberg Washington Program in Communications Policy Studies and was delivered at their forum "Global Disasters and International Information Flows," October 1986. Chapter 7 is a revised version of "The International Politics of Arctic Shipping: An American Perspective," in Franklyn Griffiths, ed.,

ACKNOWLEDGMENTS

The Politics of the Northwest Passage (Montreal: McGill-Queen's University Press, 1987), 115–133.

An earlier version of chapter 8, cast in the form of a review essay, appeared as "International Regimes: Toward a New Theory of Institutions," *World Politics* 39, no. 1 (October 1986), 104–122. Copyright © 1986 by Princeton University Press. Adapted with permission of Princeton University Press. Chapter 9 is based on a preliminary and much less substantial paper prepared for delivery at the annual meetings of the American Political Science Association in September 1983.

O.R.Y.

International Cooperation

The Problem of International Cooperation

A persistent strand of Western social thought envisions harmony as a natural outgrowth of the interactions of autonomous actors (whether individuals, corporations, or nation states). The invisible hand of perfectly competitive markets, a mechanism that is said to transform the self-interested actions of numerous buyers and sellers into collective outcomes that are efficient in allocative terms, undoubtedly constitutes a paradigm for this line of thought. But it is easy, not to mention comforting, to assume that similar processes are at work in other realms, including interactions among the members of international society. On this account, cooperation is the normal condition of human affairs. It is therefore unnecessary to embark on any elaborate or time-consuming effort to explain the occurrence of cooperation in specific social arenas or issue areas. A happy consequence of this situation, moreover, is that we need not devote any great amount of time or energy to the search for devices to overcome tendencies toward disharmony or to promote enhanced cooperation.

This line of thought is certainly appealing, suggesting as it does that it is ordinarily safe for individuals to concentrate on pursuing their own ends without worrying about the collective consequences of their behavior. Unfortunately, however, it runs directly counter to some of the most powerful and well-documented findings produced by social scientists working in a variety of fields, including international relations. As almost everyone understands by now, rational egoists making choices in the absence of effective rules or social conventions can easily fail to realize feasible joint gains, ending up with outcomes that are suboptimal (sometimes drastically suboptimal) for all parties concerned.[1] It does not even take extreme conflicts of interest, like those known to game theoreticians as zero-sum conflicts, to produce collective outcomes that are socially undesirable or, in other words, to generate collective-action problems.[2] Any situation that exhibits the characteristics of the coordination problem, a form of interdependent decision making that need not involve any conflict of interest at all, can eventuate in a collective outcome in which the participants fail to realize perfectly feasible joint gains.[3] And the probability that such suboptimal outcomes will occur generally rises as the number of participants increases. Much the same is true of mixed motive or competitive/cooperative situations that display the essential features of what game theoreticians call "chicken" or "battle of the sexes."[4] But the classic exemplar of these problems of collective action is undoubtedly the prisoner's dilemma, a theoretical construct that many analysts have scrutinized closely in abstract terms and applied repeatedly to the world of international relations as well as to numerous other substantive realms.[5] With alarming frequency, those forced to make choices in situations resembling the prisoner's dilemma experience severe difficulties in achieving the level of cooperation needed to avoid mutual losses.[6]

A similar message flows from the emerging literature on situations involving what Cross and Guyer call "inappropriate reinforcement structures [that] give rise to traps."[7] Such traps occur when initial rewards or

1. Such phenomena, grouped under the rubric of collective-action problems, constitute a principal preoccupation of the growing literature on public choice. For a theoretically sophisticated review of these problems consult Russell Hardin, *Collective Action* (Baltimore: Johns Hopkins University Press, 1982).

2. A zero-sum conflict is a situation in which the preferences of the players or participants over the set of available alternatives are strictly or diametrically opposed. Though the zero-sum condition has proven theoretically fruitful, most analysts would agree that zero-sum conflicts are the exception rather than the rule under real-world conditions. For a comprehensive discussion of zero-sum "games" see Anatol Rapoport, *Two-Person Game Theory* (Ann Arbor: University of Michigan Press, 1966).

3. Problems of this type stimulated Schelling's well-known analysis of focal points and salience. For a seminal account of the coordination problem consult Thomas C. Schelling, *The Strategy of Conflict* (Cambridge: Harvard University Press, 1960), especially chap. 4.

4. For an early but classic treatment of these situations see R. Duncan Luce and Howard Raiffa, *Games and Decisions* (New York: Wiley, 1957).

5. Consult Glenn H. Snyder and Paul Diesing, *Conflict among Nations* (Princeton: Princeton University Press, 1977); Thomas C. Schelling, *Micromotives and Macrobehavior* (New York: Norton, 1978), chap. 7; and Robert Axelrod, *The Evolution of Cooperation* (New York: Basic Books, 1984).

6. Anatol Rapoport and Albert M. Chammah, *Prisoner's Dilemma: A Study of Conflict and Cooperation* (Ann Arbor: University of Michigan Press, 1965).

7. John G. Cross and Melvin J. Guyer, *Social Traps* (Ann Arbor: University of Michigan Press, 1980), 34.

reinforcements lead to learned behavior or habits that subsequently produce" consequences that the victims would rather avoid."[8] Situations of this sort become social traps "when many victims are caught in parallel, when escapes from individual traps are influenced by social action, or when there is a great deal of interaction among victims of the same trap."[9] Traps arise in a wide range of social environments. They afflict nations caught in arms races as well as individuals locked into habits of smoking or drug addiction.

What is more, analysts have demonstrated again and again in studies of an empirical nature that the cooperation required to solve collective-action problems or to escape from social traps is elusive in the world of international relations. Wars frequently leave all the participants exhausted and faced with severe losses of welfare. Arms races commonly initiate action/reaction processes that serve only to reduce the security of all the participants.[10] Trade wars featuring successive rounds of competitive tariffs or currency devaluations generally succeed only in reducing the overall economic well-being of the members of international society.[11] Competition in the consumptive use of renewable resources, such as fish or marine mammals, often precipitates a tragedy of the commons, a condition in which everyone loses as a result of the depletion or destruction of the resource.[12] Much the same is true of the degradation of large ecosystems that results from competitive efforts to exploit these systems for the purpose of disposing of residuals or wastes.[13] Traplike behavior is equally widespread in international society. Less developed states, for example, are easily drawn into a pattern of borrowing without realizing the degree to which debt servicing will later constrain their economic options. And the short-run rewards associated with the use of renewable natural resources regularly give rise to patterns of consumption that are incompatible with conservation in that

8. Ibid., 4.

9. Ibid., 27.

10. Among students of international relations, this phenomenon has given rise to the concept of a security dilemma. See John H. Herz, *International Politics in the Atomic Age* (New York: Columbia University Press, 1959), and Robert Jervis, "Cooperation under the Security Dilemma," *World Politics* 30 (1978), 167–214.

11. For a well-known study of a classic case see Charles P. Kindleberger, *The World in Depression, 1929–1939* (Berkeley: University of California Press, 1973).

12. See also Garrett Hardin and John Baden, eds., *Managing the Commons* (San Francisco: W. H. Freeman, 1977).

13. For a range of perspectives on problems of the global commons consult Finn Sollie et. al., *The Challenge of New Territories* (Oslo: Universitetsforlaget, 1974); Seyom Brown, Nina W. Cornell, Larry L. Fabian, and Edith Brown Weiss, *Regimes for the Ocean, Outer Space, and Weather* (Washington, D.C.: Brookings, 1977); and Oran R. Young, *Resource Management at the International Level* (London and New York: Pinter and Nichols, 1977).

they conflict with the requirements of achieving sustained yields over time.

These and other similar examples are surely sufficient to demonstrate that we can no longer afford the luxury of taking harmony for granted, especially in the context of international society. On the contrary, cooperation is a striking achievement whenever and wherever it occurs, and there is every reason to believe that cooperation will become more elusive in many realms as growing human populations, enhanced capabilities, and rising expectations generate more severe conflicts of interest as well as greater demands on the earth's natural systems.[14]

Does this mean, as many students of international relations have concluded, that there is now a compelling need for the creation of a world government, a central organization or public authority capable of restricting the sovereignty of individual states and bringing pressure to bear on the members of international society to comply with rules designed to ensure the achievement of cooperation?[15] Not necessarily. By itself, the establishment of such a public authority cannot ensure the achievement of cooperation. We have known for some time that simply introducing organizational arrangements in the absence of the social conditions required to sustain cooperation is not sufficient to solve collective-action problems in any human society.[16]

Even more important, there are good grounds for concluding that the creation of a world government is not necessary to solve the problem of cooperation in international society. What this problem does suggest is that the members of international society will frequently experience powerful incentives to accept a variety of behavioral constraints in the interests of maximizing their own long-term gains, regardless of their attitudes toward the common good.[17] Whether individual actors justify their behavior in terms of rule utilitarianism, a system of ethics, or some sort of nonutilitarian contractarianism is beside the point at this juncture. The fact is that it is easy to comprehend why the members of

14. For illustrations of the growing literature on this theme consult Barbara Ward and Rene Dubos, *Only One Earth: The Care and Maintenance of a Small Planet* (New York: Norton, 1972), and Richard A. Falk, *This Endangered Planet* (New York: Random House, 1971).

15. For a range of examples of the case for world government see Emery Reves, *The Anatomy of Peace* (New York: Harper, 1945); Cord Meyer, *Peace or Anarchy* (Boston: Little, Brown, 1947); and Grenville Clark and Louis B. Sohn, *World Peace through World Law*, 2d ed. (Cambridge: Harvard University Press, 1960).

16. For an elegant discussion of this point see Hedley Bull, *The Anarchical Society: A Study of Order in World Politics* (New York: Columbia University Press, 1977).

17. See also the essays in Stephen D. Krasner ed., *International Regimes* (Ithaca: Cornell University Press, 1983), particularly Robert O. Keohane, "The Demand for Regimes," 141–171.

international society would willingly abandon a Hobbesian state of nature for a world featuring recognized social institutions, though they may show little interest in the introduction of elaborate organizations like those associated with a world government.

Starting from these premises, recent work in the field of international political economy licenses the conclusion that international regimes and, more broadly, international institutions are properly understood as responses to the pervasive collective-action problems that make cooperation problematic at the international level.[18] A rapidly growing stream of analysis explores the role of institutional arrangements in solving the problem of cooperation with particular reference to international trade and monetary issues. Yet there remain substantial ambiguities about the core meaning of the concept of an international regime as well as the relationship between regimes and international institutions more broadly.[19] What is more, there are serious questions regarding the applicability of this line of analysis to other areas of international relations. These concerns lead directly to the central themes of this book. The chapters that follow seek to place a firmer theoretical foundation under the study of international regimes by demonstrating that they belong to the larger class of social institutions and showing how our general understanding of social institutions can elucidate the conditions governing cooperation in international society. In more applied terms, the book undertakes to demonstrate the range of applicability of this conception of international regimes by using it to illuminate problems of international cooperation relating to natural resources and the environment.

Social institutions are identifiable practices consisting of recognized roles linked by clusters of rules or conventions governing relations among the occupants of these roles. We are all familiar with the roles of husband and wife in institutions of marriage, buyer and seller in exchange systems, owner and nonowner in structures of property rights, and voter and candidate in electoral systems. The rules or conventions that grow up around these roles and that constitute the superstructure of social institutions ordinarily encompass sets of rights or entitlements (for example, the use rights of property owners or the voting rights of members of the electorate) as well as sets of behavioral prescriptions (for

---

18. In addition to Krasner, *International Regimes,* see Robert O. Keohane, *After Hegemony: Cooperation and Discord in the World Political Economy* (Princeton: Princeton University Press, 1984).

19. See the criticisms expressed in Susan Strange, *"Cave! hic dragones:* A Critique of Regime Analysis," in Krasner, *International Regimes,* 337–354, and Friedrich Kratochwil, "The Force of Prescriptions," *International Organization* 38 (1984), 685–708.

example, the rules of trespass constraining the actions of nonowners of property or the rules governing the financing of campaigns for elective office).[20]

International institutions are social institutions governing the activities of the members of international society. It is easy enough to understand why the recent revival of interest in international institutions has come to focus on the specialized institutional arrangements or regimes governing international trade and monetary relations. These arrangements lend themselves to an examination of the links between politics and economics, one of the hallmarks of the international political economy approach.[21] International trade and monetary issues have given rise to unusually elaborate and intriguing institutional arrangements in modern times (for example, the postwar international monetary regimes). Yet today, these regimes are under pressure from a variety of quarters so that there is an obvious need to improve our understanding of the conditions of sustained cooperation in international trade and monetary relations. Nonetheless, collective-action problems arise regularly in many other areas of international relations. In order both to deepen our understanding of international regimes and to generalize our knowledge of these institutions, therefore, it is necessary to extend the international political economy approach into new substantive areas.[22] By focusing on regimes for natural resources and the environment, this book seeks to contribute to the fulfillment of this goal.

20. A discussion that includes some helpful observations about approaches to identifying the component elements of social institutions in empirical terms is Robert Axelrod, "An Evolutionary Approach to Norms," *American Political Science Review* 80 (1986), 1095–1111.

21. Robert Gilpin, *The Political Economy of International Relations* (Princeton: Princeton University Press, 1987).

22. For some suggestive efforts to extend this approach to the study of cooperation in security affairs see Kenneth A. Oye, ed., *Cooperation under Anarchy* (Princeton: Princeton University Press, 1986), pt. 2.

PART ONE

INTERNATIONAL REGIMES
IN THEORY

Prologue

Much of the growing literature on international regimes consists of descriptive accounts of specific institutional arrangements and fails to address the underlying analytical questions associated with the use of the concept of regimes as an organizing principle in the study of recurrent international phenomena. As a result, the whole enterprise of regime analysis continues to rest on a shaky foundation. The concept of a regime itself is often used so loosely that critics have reasonably questioned whether the concept is anything but a woolly notion likely to produce more confusion than illumination. Furthermore, few sustained efforts have been made to provide a compelling account of the degree to which institutional arrangements operate to determine collective outcomes in international society. It is therefore hard to judge whether the study of regimes deserves to become a permanent feature of research in international relations.

For this reason, the chapters of Part 1 focus on international regimes in theory. Chapter 1 elaborates on the proposition that international regimes are social institutions. In the process, it develops some generic ideas regarding rights and rules in international relations. Chapter 2, which centers on an extensive discussion of the distinction between institutions and organizations at the international level, is critical to differentiating this study from the more conventional literature on international organizations. While the use of the terms *institution* and *organization* is of no great consequence, the distinction between the two is what makes it possible to use the study of international regimes as a vehicle for deepening our understanding of the prospects for cooperation in international society. Chapter 3 turns to the question of how institutional arrangements serve not only to solve collective-action problems but also to structure the character of collective outcomes at the international level. The argument is pivotal in that it indicates why institutional arrangements are important as independent variables in our efforts to explain the variance in collective outcomes in international society. In this context, chapter 4, which deals with the formation and transformation of international regimes, takes on added significance. To the extent that institutions matter as determinants of collective outcomes, our interest in exploring the dynamics of institutional arrangements is heightened.

Overall, the four chapters of Part 1 place the analysis of international institutions in general and international regimes in particular squarely within the intellectual tradition of studies of social institutions. Certainly, no one would argue that the study of social institutions constitutes a unified and theoretically sophisticated field of enquiry. There is too much diversity in both definitional matters and theories of causation to allow us to speak of an acknowledged paradigm for the study of social institutions. Even so, placing international regimes within the larger domain of social institutions provides access to a significant fund of intellectual capital that we can readily tap in our efforts to comprehend a range of international phenomena. In the process, it should help us to transcend sterile definitional arguments and to move toward examining an array of interesting substantive issues suggested by a focus on international regimes.

CHAPTER ONE

International Regimes:
An Institutional Perspective

We live in a world of international regimes. Some of them deal with monetary issues (for example, the Bretton Woods system and its successors); others govern international trade in commodities (for example, the coffee agreement). Some regimes serve to direct the use of natural resources at the international level (for example, the international arrangements for whaling) or to advance the cause of conservation (for example, the agreement on polar bears). Still other regimes address problems pertaining to the control of armaments at the international level (for example, the nuclear nonproliferation regime) or the management of power in international society (for example, the neutralization agreement for Switzerland). And there are some international regimes that encompass several issues within well-defined geographical areas (for example, the Svalbard regime and the Antarctic Treaty System).

International regimes cover a wide spectrum in terms of functional scope, geographical domain, and membership. Functionally, they range from the narrow purview of the polar bear agreement to the broad concerns of the arrangements for Antarctica and outer space. The geographical area covered may be as limited as the highly restricted domain of the regime for fur seals in the North Pacific or as far-flung as that of the global regimes for international air transport (the ICAO/IATA system) or for the control of nuclear testing. With respect to membership, the range runs from two or three members, as in the regime for high-seas fishing established under the International North Pacific Fisheries Convention, to well over a hundred members, as in the nuclear nonproliferation regime. What is most striking, however, is the sheer number of international regimes. Far from being unusual, they are common throughout international society. *And, yet, they are not generally part of planning, practice/discourse*

It is therefore surprising that it took so long for students of international affairs to focus intensively on international regimes and that much of the newly emerging literature on regimes is weak, particularly in analytic terms. The last decade has brought a surge of interest in the study of international regimes.[1] We now have fairly extensive descriptive accounts of some specific regimes[2] and some speculative ideas about phenomena such as regime change.[3] Even so, the fundamental character of international regimes remains elusive, and there is nothing approaching consensus on the role of regimes in international society. Considering the pervasiveness of regimes at the international level, the resultant limits on the state of our knowledge of international regimes constitute a serious deficiency. In this chapter, I take some preliminary steps toward filling this gap, laying out a conceptual framework to be used in studying international regimes systematically.[4]

The Core Concept

Regimes are social institutions governing the actions of those involved in specifiable activities or sets of activities. Like all social institutions, they are practices consisting of recognized roles linked together by clusters of

1. See, among others, Richard N. Cooper, "Prolegomena to the Choice of an International Monetary System," *International Organization* 29 (1975), 63–97; Ernst B. Haas, "On Systems and International Regimes," *World Politics* (1975), 147–174; John Gerard Ruggie and Ernst B. Haas, eds., *International Responses to Technology,* special issue of *International Organization* 29 (1975); Robert O. Keohane and Joseph S. Nye, *Power and Interdependence* (Boston: Little, Brown, 1977); Oran R. Young, *Resource Management at the International Level: The Case of the North Pacific* (London and New York: Pinter and Nichols, 1977); Seyom Brown, Nina W. Cornell, Larry L. Fabian, and Edith Brown Weiss. *Regimes for the Ocean, Outer Space, and Weather* (Washington, D.C.: Brookings, 1977); and Stephen D. Krasner, ed., *International Regimes* (Ithaca: Cornell University Press, 1983). In addition, international lawyers have talked about international regimes for some time. See L. F. E. Goldie, "The Management of Ocean Resources: Regimes for Structuring the Maritime Environment," 155–247 in Cyril E. Black and Richard A. Falk, eds., *The Future of the International Legal Order,* 4: *The Structure of the International Environment* (Princeton: Princeton University Press, 1972).

2. For a range of examples see William M. Ross, *Oil Pollution as an International Problem: A Study of Puget Sound and the Strait of Georgia* (Seattle: University of Washington Press, 1973); M. M. Sibthorp, ed., *The North Sea: Challenge and Opportunity* (London: Europa, 1975); Kenneth Dam, *Oil Resources* (Chicago: University of Chicago Press, 1976); and Arild Underdal, "The Politics of International Fisheries Management," unpublished paper, Oslo, 1978.

3. For example, Keohane and Nye, *Power and Interdependence,* pt. 2, and Robert O. Keohane, *After Hegemony: Cooperation and Discord in the World Political Economy* (Princeton: Princeton University Press, 1984).

4. See also Oran R. Young, *Resource Regimes: Natural Resources and Social Institutions* (Berkeley: University of California Press, 1982).

rules or conventions governing relations among the occupants of these roles. It is important not to conflate institutions and functions, though the operation of institutional arrangements can and frequently does contribute to the fulfillment of certain functions. Like other social in- stitutions, regimes may be more or less formally articulated, and they may or may not be accompanied by explicit organizations.[5]

Though students of international affairs are often rather imprecise in their usage, they generally find it helpful to divide the category of international institutions into two more or less distinct subsets: interna- tional orders and international regimes. International orders are broad, framework arrangements governing the activities of all (or almost all) the members of international society over a wide range of specific issues. We speak of the international political order, for example, as a system of territorially based and sovereign states that interact with one another in the absence of any central government. Similarly, we think of the inter- national economic order as a system of exchange relationships in which buyers and sellers throughout international society are free to partici- pate in an array of international markets. As such, this order subsumes a collection of more specific arrangements, like the international trade regime, the international monetary regime, and the international re- gime pertaining to direct foreign investment. When Third World states call for a new international economic order, therefore, they are asking for fundamental revisions in the basic framework of economic institu- tions prevailing in international society; they are not just advocating certain alterations in the specific arrangements governing trade or mon- etary matters.

International regimes, by contrast, are more specialized arrange- ments that pertain to well-defined activities, resources, or geographical areas and often involve only some subset of the members of interna- tional society. Thus, we speak of the international regimes for whaling, the conservation of polar bears, the use of the electromagnetic spec- trum, and human activities in Antarctica. Because the members of inter- national society are states, the rules or conventions of regimes apply, in the first instance, to the actions of states. Yet the parties actually engag- ing in the activities governed by regimes are frequently private entities such as multinational corporations, banks, fishing companies, or pri-

---

5. Even when a regime is articulated formally in a contract or a treaty, informal rules typically grow up in conjunction with the resulting institutional arrangement in practice. This fact suggests the importance of a behavioral approach to the empirical identification of regimes. For some similar observations regarding the study of norms see Robert Axelrod, "An Evolutionary Approach to Norms," *American Political Science Review* 80 (1986), 1096–1098.

vately owned airlines. States participating in international regimes must therefore assume some responsibility for ensuring that these parties comply with the dictates of the regimes. It follows that implementing the terms of international regimes commonly involves a two-step procedure, a phenomenon that is less characteristic of institutional arrangements at the domestic level.[6]

Specific regimes are regularly nested in international orders in the sense that they build on the foundation provided by more general institutions rather than offer arrangements that are unconnected to broader orders or that even conflict with the provisions of such institutions. The commodity regimes dealing with products such as coffee, sugar, or tin are meant to supplement the market mechanisms of the international economic order rather than to supplant them.[7] The specialized arrangements now emerging to cope with the problem of ozone depletion incorporate the more general arrangements that govern interstate relations regarding environmental matters; they are not intended to replace these arrangements. Similarly, the nuclear nonproliferation regime operates within the prevailing structure of security arrangements in which relationships of deterrence constitute a key element in the interactions among states.[8] It follows, therefore, that analyses of specific international regimes cannot be complete without some consideration of the extent to which they are embedded in the larger institutional orders that are operative in international society.

The mere existence of a social institution will lend an element of orderliness to the activities it governs. But there is no reason to assume that institutional arrangements will guide human activities toward well-defined substantive goals, such as enduring peace, economic efficiency, or optimum yields from renewable resources.[9] The more specific concept of an international regime contains no intrinsic metaphysical or teleological orientation, though those involved in the creation or reform of any given regime often attempt to shape its contents with clear-cut

6. Among other things, this fact explains why those endeavoring to elicit compliance with the rights and rules of international regimes often turn to domestic courts. See Richard A. Falk, *The Role of Domestic Courts in the International Legal Order* (Syracuse: Syracuse University Press, 1964).

7. See also Mark W. Zacher, "Trade Gaps, Analytical Gaps: Regime Analysis and International Commodity Trade Regulation," *International Organization* 41 (1987), 173–202.

8. Roger K. Smith, "Explaining the Non-Proliferation Regime: Anomalies for Contemporary International Relations Theory," *International Organization* 41 (1987), 253–281.

9. In fact, institutional arrangements can produce consequences that are widely viewed as bad or undesirable in social terms. Consider, for example, institutions such as slavery in domestic societies or war in international society.

objectives in mind.[10] It is, however, possible to identify several key components that every regime possesses.

The Substantive Component

The core of every international regime is a cluster of rights and rules. Though they may be more or less extensive or formally articulated, some such institutional arrangements structure the opportunities available to actors, and their exact content is a matter of intense interest to these actors.[11]

A right is anything to which an actor (individual or otherwise) is entitled by virtue of occupying a recognized role. The role of human being, for example, is often said to carry with it a right to life. In the United States, the role of citizen carries with it the right to vote in elections, the right to speak freely, and the right to move about at will. We are now witnessing vigorous campaigns to clarify and, in some cases, to redefine the rights of women, children, hospital patients, inmates in prisons, and animals. Roles commonly carry with them bundles of rights that may be more or less extensive and whose precise content is subject to change over time.[12] Of course, the possession of a right does not guarantee that an actor will actually receive those things to which he is entitled under the terms of the right. Although rights are often respected, even acknowledged rights are violated with considerable frequency in real-world social contexts.

Several categories of rights figure prominently in international regimes. Property rights may take the form of private-property rights (for example, rights to commodities traded internationally) or common-property rights (for example, rights to air space or deep-seabed min-

10. In real-world situations, those involved in regime formation virtually never operate behind a Rawlsian "veil of ignorance." See John Rawls, *A Theory of Justice* (Cambridge: Harvard University Press, 1971), chap. 3.

11. Though this is not the place to address problems of operationalization in detail, a few comments on the empirical analysis of rights and rules are in order here. Empirically, rights and rules can be approached either in terms of the expectations shared by the members of a given group or in terms of observable regularities in the behavior of the group's members. As the two approaches are not mutually exclusive, it may well be instructive to make use of both in the effort to pin down rights and rules empirically. This suggests that there is no necessary connection between the rights and rules actually operative in a social environment and the statements regarding rights and rules contained in formal documents such as treaties or conventions, though it may still prove useful to turn to such documents for an easily accessible approximation of the substantive component of specific institutional arrangements. For some additional observations pertaining to these issues see Axelrod, "An Evolutionary Approach to Norms."

12. On the idea of bundles of rights see Charles A. Reich, "The New Property," *Yale Law Journal* 73 (1964), 733–787.

erals).[13] Because of the prevalence of common-property arrangements at the international level, international regimes often focus on the development of use-and-enjoyment rights. These may be exclusive in nature (for example, the right to exploit a given tract on the deep seabed) or they may be formulated in nonexclusive terms (for example, the right to use certain international straits).[14] But all such rights are designed to ensure the availability of key resources to actors under conditions in which private-property arrangements are infeasible or undesirable. International regimes may also encompass an assortment of other rights, including the right to protection against certain forms of aggression, the right to receive specified benefits from international transactions or productive operations, the right to trade on favorable terms with other members of international society, and the right to participate in making collective decisions under the terms of a given regime.

In contrast to rights, rules are well-defined guides to action or standards setting forth actions that members are expected to perform (or to refrain from performing) under appropriate circumstances.[15] Any given rule exhibits the following features: an indication of the relevant subject group, a behavioral prescription, and a specification of the circumstances under which the rule is operative. In some societies, there are near-universal rules enjoining individuals to tell the truth and to keep promises in their dealings with other members of the society. But rules may be directed toward some clearly designated group, as in the case of ethical prescriptions governing the behavior of teachers, doctors, or lawyers. Or rules may focus on some specific activity, as in the case of prescriptions pertaining to civil aviation or maritime commerce. Of course, the existence of an acknowledged rule does not guarantee that the members of the subject group will always comply with its requirements. Even in well-ordered societies, noncompliance with rules is a common occurrence.

Among the numerous rules associated with international regimes, three general categories are particularly prominent. First, there are use rules. For example, members of the ICAO/IATA system are required to

13. Consult, among others, Eirik Furubotn and Svetozar Pejovich, "Property Rights and Economic Theory: A Survey of Recent Literature," *Journal of Economic Literature* 10 (1972), 1137–1162.

14. The result may be described as a system of "restricted" common property: see J. H. Dales, *Pollution, Property, and Prices* (Toronto: University of Toronto Press, 1968), 61–65.

15. My use of the concept "rules" differs somewhat from that prevalent in recent contributions to jurisprudence. Compare H. L. A. Hart, *The Concept of Law* (Oxford: Oxford University Press, 1961), and Ronald Dworkin, *Taking Rights Seriously* (Cambridge: Harvard University Press, 1977), especially chaps. 2 and 3.

follow certain safety rules in using international airspace; those engaged in fishing operations are usually subject to rules pertaining to the conservation of fish stocks; and those using international shipping lanes are subject to rules designed to maximize safety and minimize marine pollution. Frequently, such use rules take the form of limitations on the exercise of rights. Just as rights commonly safeguard the freedom of actors to behave in certain ways, rules often spell out restrictions on the freedom of actors to do as they wish.[16] Liability rules constitute a second category. They spell out the locus and extent of responsibility in cases of (usually unintended) injury to others arising from the actions of individual parties under the terms of a regime. They range from rules concerning compensation for expropriation of foreign investments under various circumstances to rules pertaining to responsibility for cleaning up marine environments in the wake of accidents.[17] Finally, international regimes often specify a variety of procedural rules, which deal with the handling of disputes or the operation of explicit organizations associated with regimes.

At the domestic level, collections of rights and rules are commonly supplemented by extensive sets of regulations and incentive systems. Regulations are administrative directives promulgated by public authorities and specifying conditions under which certain actors are to operate on a day-to-day basis. They are widely used to translate rights and rules formulated in general terms into working managerial arrangements applicable to real-world situations.[18] Incentive systems, on the other hand, are penalties and rewards employed by public authorities for the purpose of altering the behavior of actors in desired directions.[19] Perhaps the classic cases of incentive systems are taxes and subsidies.

Obviously, regulations and incentive systems are used less extensively in international regimes than in institutional arrangements operating at the national and subnational levels. They require a public agency possessing a measure of authority and power, and such agencies are far less characteristic of highly decentralized social systems like international

---

16. See G. H. von Wright, *Norms and Actions* (New York: Humanities Press, 1963).

17. On liability rules and their significance compare R. H. Coase, "The Problem of Social Cost," *Journal of Law and Economics* 3 (1960), 1–44, and Guido Calabresi and A. Douglas Melamed, "Property Rules, Liability Rules, and Inalienability: One View of the Cathedral," *Harvard Law Review* 85 (1972), 1089–1128.

18. I use the term *regulation* in a sense somewhat different from that used in discussions of public regulation of private industries. For a clear example of this usage see George Stigler, *The Citizen and the State: Essays on Regulation* (Chicago: University of Chicago Press, 1975).

19. Incentive systems can also be used to raise or disburse revenues. Their primary purpose, however, is to structure the behavior of identifiable groups of actors.

society than of the more centralized systems that are common at the national level. Yet international regimes accompanied by explicit organizations can and sometimes do employ these devices. The International Monetary Fund, for example, has promulgated extensive regulations pertaining to the drawing rights of individual members, and the proposed International Seabed Authority would be able to regulate production of manganese nodules to implement more general rules concerning such matters as the impact of deep-seabed mining on the world nickel market.[20]

The Procedural Component

A procedural component of international regimes encompasses recognized practices for handling situations requiring social or collective choices. Situations of this type arise whenever it is necessary or desirable to aggregate the (nonidentical) preferences of two or more individual actors into a group choice.[21] Such problems occur in most social systems; they range from the selection of individuals to fill certain positions to the establishment of the terms of trade in exchange relationships and decisions on the distribution of valued goods and services.

Several types of social-choice problems arise regularly in international regimes.[22] Some involve the allocation of factors of production (for example, deep-seabed mining tracts, total allowable catches in fisheries, segments of the global electromagnetic spectrum). Such problems are especially difficult to handle at the international level owing to the prevalence of common rather than private property. Other social-choice problems relate to issues with explicit distributive implications (for example, adjustments of exchange rates or the disposition of economic returns or rents associated with deep-seabed mining). Collective choices are also required to settle disputes. Typically, these disputes arise from efforts to apply general rights and rules to the complexities of real-world situations. There are cases, as well, in which collective decisions are necessary to determine the sorts of activities to permit in an area like Antarctica, to resolve conflicts between competing uses of the same resource, and to organize collective sanctions aimed at obtaining compliance with the rights and rules of an international regime.

20. See, for example, Robert Z. Aliber, *The International Money Game* (New York: Basic Books, 1976).
21. For a general analysis of social choice see A. K. Sen, *Collective Choice and Social Welfare* (San Francisco: W. H. Freeman, 1970).
22. Problems of social choice pertaining to the selection and reform of international regimes per se are discussed in chapters 4 and 9 below.

Social-choice mechanisms are institutional arrangements specialized to the resolution of problems of social choice arising within the context of particular regimes. Like other components of regimes, these mechanisms may be more or less formalized, and it is typical for a regime to make use of several at the same time. The range of these mechanisms is wide, encompassing such devices as the principle of first come, first served, markets, voting systems, bargaining, administrative decision making, adjudication, unilateral action backed by coercion, and organized violence.[23] Certain conditions are required for the effective operation of each of these mechanisms; we may therefore assume that individual mechanisms will be associated with particular types of social systems. The most striking features of international society in this regard are its relatively small number of formal members and its high level of decentralization of authority. The social-choice mechanisms characteristic of such a social setting are the principle of first come, first served, bargaining, various forms of coercion, and, to a lesser degree, markets. We should consequently expect problems of social choice in international regimes to be handled through these procedures.[24] Still, voting and administrative decision making are not altogether absent from international regimes. Voting is significant, for example, in cases like the international monetary regime and the ICAO/ IATA system. But there can be no doubt that unilateral claims, bargaining, and coercion are central to the processes of arriving at social choices within most international regimes.

It is also worth noting that some regimes do not possess social-choice mechanisms of their own. They may rely on the institutional arrangements of some encompassing international order or share mechanisms with other regimes in handling problems of social choice. Such situations are common where adjudication or voting is employed in reaching collective choices. For example, the same courts may resolve conflicts of interest pertaining to civil liberties, business activities, and land use. In principle, the International Court of Justice or the General Assembly of the United Nations could be employed to deal with many of the social-choice problems arising under specific international regimes. In practice, however, various combinations of bargaining and coercion geared to the problems of specific regimes constitute the norm at the international level.

23. The classic study, which focuses on voting systems, is Kenneth Arrow, *Social Choice and Individual Values*, 2d ed. (New York: Wiley, 1963).

24. Oran R. Young, "Anarchy and Social Choice: Reflections on the International Polity," *World Politics* 30 (1978), 241–263.

Implementation

Smoothly functioning international regimes are not easy to establish.[25] Rights are not always respected, and even widely acknowledged rules are violated with some frequency. Nor is it reasonable to assume that participants will always accept the results generated by social-choice mechanisms as authoritative and abide by them. Accordingly, it is important to think about the effectiveness of international regimes,[26] and this observation suggests an examination of compliance mechanisms as a third major component of these regimes.

Any discussion of compliance must deal with the incentives of those who are parties to institutional arrangements. What are the benefits and costs of complying with rights and rules, in contrast to violating them? How do individual actors decide whether to comply with the substantive provisions of international regimes? There is a tendency to assume that the typical actor will violate such provisions so long as the probability of being caught in specific instances is low—a line of reasoning that implies that the existence of effective enforcement procedures is essential to the achievement of compliance. But this argument appears to be quite wide of the mark in many real-world situations. It is not difficult to identify a variety of factors, other than detection and the imposition of public sanctions, that exert effective pressure for compliance, especially in conjunction with long-run perspectives on iterative behavior. There is also no basis for assuming that individual actors make large numbers of discrete benefit/cost calculations regarding compliance with the provisions of international regimes. Rather, they commonly develop rules or policies in this realm, and it seems reasonable to expect that long-term socialization as well as feelings of obligation will play a part in the articulation of these policies.[27]

A compliance mechanism is any institution or set of institutions publicly authorized to promote compliance with the substantive provisions of a regime or with the outcomes generated by its social-choice mechanisms. This characterization conjures up an image of formal governmental agencies, and such agencies undoubtedly do constitute the classic means of coping with compliance problems. But less formal compliance mechanisms are common, and highly decentralized social systems, like

25. That is, reality seldom approximates the condition of "perfect compliance" discussed in Rawls, *A Theory of Justice*, 351.
26. For a similar observation about institutional arrangements at the domestic level see A. Myrick Freeman, "Environmental Management as a Regulatory Process," Resources for the Future Discussion Paper D–4, (Washington, D.C., January 1977).
27. For an intriguing empirical example see Abram Chayes, "An Enquiry into the Workings of Arms Control Agreements," *Harvard Law Review* 85 (1975), 905–969.

international society, typically rely on them.[28] The result is apt to be a heavy emphasis on self-interest calculations coupled with publicly recognized procedures for self-help to redress wrongs.[29] Alternatively, such conditions sometimes give rise to a reliance on arrangements in which supranational agencies are employed to gather information and to inspect the activities of individual actors, but decentralized procedures are retained for the application of sanctions (for example, the nuclear nonproliferation regime, the ICAO/IATA system, and many of the regional-fisheries regimes).[30]

From the point of view of the members of a regime, the development of compliance mechanisms poses an investment problem. Any expenditure of resources on such mechanisms will generate opportunity costs, and declining marginal returns from such investments virtually always become pronounced before perfect compliance is reached. Accordingly, it is safe to assume that the members of a regime will rarely attempt to devise compliance mechanisms capable of eliminating violations altogether. Exactly where equilibrium will occur with respect to these investment decisions depends on the assumptions made about the members of international regimes. Given the dispersal of responsibility that goes with the decentralization of authority in international society, however, it is safe to conclude that underinvestment in compliance mechanisms is characteristic of international regimes.[31] Still, various types of compliance mechanisms do operate at the international level,[32] and such mechanisms must be treated as a third major component of international regimes.

Clarifying Observations

It is possible to argue that some regime must be present in every international activity: regimes can vary greatly in extent, and extreme

28. Oran R. Young, *Compliance and Public Authority: A Theory with International Applications* (Baltimore: Johns Hopkins University Press, 1979), especially chaps. 4 and 5.

29. For empirical examples consult Lucy Mair, *Primitive Government* (Bloomington: Indiana University Press, 1977), especially chap. 1.

30. Ronald S. Tauber, "The Enforcement of IATA Agreements," *Harvard International Law Journal* 10 (1969), 1–33.

31. International regimes, like other social institutions, will ordinarily exhibit the attributes of collective goods (that is, nonexcludability and jointness of supply) to a relatively high degree. For a classic account of the problems of supplying collective goods see Mancur Olson, Jr., *The Logic of Collective Action* (Cambridge: Harvard University Press, 1965).

32. For a variety of examples see William T. Burke, Richard Legatski, and William W. Woodhead, *National and International Law Enforcement in the Ocean* (Seattle: University of Washington Press, 1975).

cases can simply be treated as null regimes. The arrangements for high-seas fishing prevailing prior to World War II, for instance, might be described as a regime based on unrestricted common property and the procedure known as the "law of capture," rather than as a situation lacking any operative regime.[33] But this line of reasoning is seriously flawed. Some activities arise de novo in the absence of prior experience (for example, international satellite broadcasting or deep-seabed mining). In such cases, we would have to develop some fiction about latent or tacit regimes to avoid the conclusion that there are situations in which no regime is present. Further, existing regimes sometimes break down, leaving a confused and inchoate situation (for example, international trade and finance in the aftermath of the Great Depression and World War II).[34] Here, too, the concept would have to be stretched excessively to assert the continued existence of a regime. What is more, avoiding the temptation to assume the presence of some regime in every specifiable activity will facilitate later discussions of the origins of regimes and of regime transformation.

In analyzing international regimes, there is also a tendency to focus on highly coherent and internally consistent constructs. Yet real-world regimes are often unsystematic and ambiguous, incorporating elements derived from several analytic constructs or ideal types. This is sometimes the result of misunderstandings on the part of those who make decisions about the creation of regimes. Much of the ambiguity, however, arises from two other factors. The development of an international regime frequently involves intense bargaining that leads to critical compromises among the interested parties. The hard bargaining that characterized the hammering out of a regime to govern deep-seabed mining illustrates this phenomenon. Additionally, international regimes generally evolve and change over time in response to various economic and political pressures. This is true even of regimes formulated initially in some comprehensive constitutional contract. With the passage of time, regimes generally acquire new features and become less consistent internally. The point of these remarks is neither to criticize existing regimes nor to argue that ideal types are unimportant in examinations of the development of international institutions. But a failure to bear in mind the distinction between ideal types and reality is bound to lead to confusion.[35]

33. Francis T. Christy and Anthony Scott, *The Common Wealth in Ocean Fisheries* (Baltimore: Johns Hopkins University Press, 1965).

34. Charles P. Kindleberger, *The World in Depression, 1929–1939* (Berkeley: University of California Press, 1973).

35. On the relationship between ideal types and reality, with special reference to the

Finally, there is an important distinction between the conditions required for the effective operation of an international regime and the consequences resulting from its operation. To illustrate, consider a regime governing international trade in some commodity based on private-property rights and a competitive market. The conditions necessary to ensure effective operation of such a regime include the availability of information about potential trades, a willingness to accept terms of trade dictated by the market, and an absence of natural monopolies.[36] The consequences of the operation of the regime, by contrast, relate to the extent to which it yields economically efficient outcomes; the degree to which it produces social costs or neighborhood effects, the attractiveness of the results in distributive terms, and so forth. Both the conditions for operation and the consequences of operation are central issues in the analysis of international regimes. But it is important to differentiate between them clearly, as well as to bear in mind that both these issues are separable from efforts to characterize the institutional content of an international regime.

REGIMES IN OPERATION

We turn now to some prominent features of international regimes as they occur under real-world conditions.

Varieties of Regimes

Variation in extent, formalization, direction, and coherence is a prominent feature of international regimes. Sometimes variations are outgrowths of underlying philosophical differences. For example, regimes resting on socialist premises encompass more extensive collections of rules as well as more explicit efforts to direct behavior toward the achievement of goals than laissez-faire regimes that emphasize decentralized decision making and autonomy for individual actors. In other cases, variation arises from the character of specific bargains struck in the process of setting up regimes, or from particular patterns of institutional evolution over time.

The extent of a regime is a matter of the number and restrictiveness of

---

theory of games, see Anatol Rapoport, *Two-Person Game Theory* (Ann Arbor: University of Michigan Press, 1966), 186–214.

36. For a succinct and lucid account of such conditions see Robert Haveman, *The Economics of the Public Sector* (New York: Wiley, 1976), 22–27.

its rights and rules. At one extreme is the case of unlimited laissez-faire, in which actors are completely free to do as they please without even the constraints of a system of property or use rights.[37] At the other extreme are arrangements featuring central planning and pervasive rules governing the actions of individual members. Between these extremes lies a wide range of mixed cases that are differentiable in terms of the extent to which they include rights and rules restricting the autonomy of individual members. Though international regimes tend to be less restrictive than institutional arrangements in domestic society, they rarely approximate the extreme of unlimited laissez-faire.

Some writers have fallen into the habit of equating regimes with the agreements in terms of which the regimes are often expressed or codified.[38] In practice, however, international regimes vary greatly in the extent to which they are expressed in formal agreements, treaties, or conventions. The current regime for Antarctica is formalized to a far greater degree than the neutralization arrangements for Switzerland. As in domestic society, moreover, it is common for informal understandings to arise within the framework established by the formal structure of an international regime. Such understandings may serve either to provide interpretations of ambiguous aspects of the formal arrangements (for example, the notion of optimum yield in conjunction with marine fisheries) or to supplement formal arrangements by dealing with issues they fail to cover (for example, the treatment of nuclear technology under the terms of the partial nuclear test-ban regime). Though it may be helpful, formalization is clearly not a necessary condition for the effective operation of international regimes.[39] There are informal regimes that have been generally successful, and there are formal arrangements that have produced unimpressive results (for example, several of the commodity agreements).[40]

Regimes are directed to the extent that they exert pressure on their members to act in conformity with some clear-cut social or collective

37. As an extreme type, this category is empirically empty. But a regime for some natural resource with no private property rights, no liability rules, and allocation based on the principle known as the "law of capture" would approach the extreme case.

38. See, for example, Arthur A. Stein, "Coordination and Collaboration: Regimes in an Anarchic World," 115–140 in Krasner, *International Regimes*.

39. As well, formalization is not a sufficient condition for the effective operation of international regimes. In this connection, note also that any definitional convention that equates regimes with the existence of formalized agreements cuts off efforts to analyze relationships betwen institutional arrangements and the formalization of such arrangements in treaties or conventions.

40. United Nations, *International Compensation for Fluctuations in Commodity Trade* (New York; 1961), and Zacher, "Trade Gaps."

goal. Various goals are feasible—including economic efficiency, distributive justice, the preservation of ecosystems, and so forth. Even where there is agreement in principle about the pursuit of some social goal, however, it may prove difficult to achieve the desired results under real-world conditions. The goal of maximum sustained yield (much less optimum yield) with respect to the marine fisheries, for example, is notoriously difficult to achieve in reality.[41] Additionally, when a regime is directed toward the achievement of several goals at once, close attention must be paid to the determination of trade-offs among these goals.[42] In the absence of systematic efforts to construct trade-off functions, any apparent directedness of a regime encompassing two or more distinct goals will be illusory.

Coherence refers to the degree to which the elements of an international regime are internally consistent. Severe internal contradictions are common in real-world regimes, even in cases where such arrangements are articulated in constitutional contracts. For example, there are often contradictions between use rights for those desiring to exploit marine resources and rights vested in adjacent coastal states to exclude outsiders. Similarly, conflicts regularly arise between the alleged requirements of indivisible state sovereignty and the obligations imposed by the rules of international regimes. It is not hard to account for these elements of incoherence in terms of the compromises necessary to achieve initial acceptance of regimes or in terms of the piecemeal evolution of regimes over time in response to changing political, economic, and social forces. But the widespread occurrence of incoherence means that we must beware of relying too heavily on neat analytic constructs in interpreting real-world situations and that we must learn to cope with the existence of contradictions.

Organizations

Though all regimes, even highly decentralized private-enterprise arrangements, are social institutions, they need not be accompanied by organizations possessing their own personnel, budgets, physical facilities, and so forth. Effective institutional arrangements lacking organizations to administer them are widespread in "primitive" societies,[43] but

---

41. P. A. Larkin, "An Epitaph for the Concept of Maximum Sustained Yield," *Transactions of the American Fisheries Society* 106 (1977), 1–11.

42. On economic approaches to such trade-offs see Richard Zeckhauser and Elmer Shaefer, "Public Policy and Normative Economic Theory," 27–101 in Raymond A. Bauer and Kenneth J. Gergen, eds., *The Study of Policy Formation* (New York: Free Press, 1968).

43. Mair, *Primitive Government*.

they are by no means confined to societies of this type. For example, free-enterprise systems relying on competitive markets are classic cases of social institutions performing vital functions in society with little in the way of administering organizations.[44] Many other social institutions—such as those governing manners, dress, and intergenerational relations—serve to structure behavior effectively with little need for organizations. Although it is undoubtedly true that international regimes characteristically involve fewer organizational arrangements than their domestic counterparts, it is important not to carry this generalization too far. The organizations associated with the international monetary regime are certainly not trivial, and the organizational arrangements contemplated in conjunction with the deep-seabed mining regime are quite complex.

Even where the need for organizational arrangements is apparent, regimes may make use of organizations created for other purposes or associated with a more comprehensive international order in preference to creating autonomous arrangements of their own. Such situations are common at the domestic level; regimes regularly leave tasks involving information collection, inspection, dispute resolution, and enforcement to agencies specializing in these matters, so that they do not require archives, court systems, or police forces of their own. At the international level, this practice appears to be less common. Situations in which regimes covering well-defined activities could benefit from such arrangements occur frequently enough. For example, it would be a distinct help in connection with the development of arms-control regimes to be able to rely on the monitoring and verification capabilities of some broader public authority.[45] And some experiments with the use of outside organizational arrangements have occurred, as exemplified by the role of the International Union for the Conservation of Nature and Natural Resources and the World Wildlife Fund in administering the regime for endangered species articulated in the Convention on International Trade in Endangered Species of Fauna and Flora.[46] But comprehensive organizational capabilities are comparatively underdeveloped in international society. Thus, the United Nations is hardly capable of inspecting activities carried out under the regime for Antarctica or resolving disputes pertaining to deep-seabed mining. It follows that

44. See Haveman, *Economics of the Public Sector*, 21, for a description of markets in precisely these terms.

45. See Richard A. Falk and Richard Barnet, eds., *Security in Disarmament* (Princeton: Princeton University Press, 1965).

46. Simon Lyster, *International Wildlife Law* (Cambridge: Grotius Publications, 1985), chap. 12.

international regimes are less tightly linked together than domestic regimes, even though they often lack extensive organizational arrangements of their own.

The emergence of organizations associated with international regimes raises a set of classic questions that are just as pressing at the international level as they are at the domestic level. How much autonomy vis-à-vis other centers of authority in the social system should these organizations possess? What sorts of decision rules should the organizations employ? How much discretion should the organizations have to make changes affecting the substantive content of the regime itself? How should the organizations be financed: where should their revenues come from and how should they be raised? How should the organizations be staffed? What sorts of physical facilities should the organizations have and where should these facilities be located? The answers to all these questions can influence the impact a regime has on its members. It is therefore to be expected that such issues will be contested vigorously, not only at the outset but also during the whole period over which the regime is effective. To the extent that administering organizations are less important in connection with international regimes than they are in domestic regimes, this sort of contention will be less pervasive at the international level. Nevertheless, it is impossible to comprehend recent negotiations relating to international monetary arrangements, deep-seabed mining, or the allocation of the broadcast frequency spectrum without paying careful attention to these questions of organizational design.[47]

Policy Instruments

Policy instruments are devices subject to deliberate or planned manipulation in the interests of achieving social goals. Such instruments can operate at different levels of generality. Changes in bundles of property rights, the promulgation of restrictive regulations, and decisions regarding individual applications for loans or mining licenses may all be treated as matters involving the use of policy instruments, but they obviously address problems occurring at different levels of generality. Policy instruments are apt to be articulated in terms that are specific to individual regimes or types of regimes. For instance, in fisheries regimes the determination of allowable catches and decisions concerning the

---

47. See, for example, Brown et al., *Regimes for the Ocean,* and Michael Hardy, "The Implications of Alternative Solutions for Regulating the Exploitation of Seabed Minerals," *International Organization* 31 (1977), 313–342.

opening and closing of harvest areas are standard issues involving policy instruments. Adjustments of exchange rates or the issuance of broadcast licenses are common policy instruments in other regimes.[48]

At the international level, a key consideration concerns the extent to which the use of policy instruments requires the existence of organizations. It is possible, for example, to redefine the contents of collections of rights and rules at occasional assemblies of the members of a regime; it may even be possible to do so as a result of unilateral actions on the part of some members of a regime to which others subsequently conform on a de facto basis. Policy instruments of this sort have an obvious appeal in highly decentralized settings like international society. This appeal may account for the current tendency to respond to problems relating to international maritime regimes by redrawing jurisdictional boundaries (that is, shifting areas or activities from the domain of international common property toward the domain of national public property) rather than by agreeing to well-defined rules for the use of common property at the international level.[49] States can pursue jurisdictional changes unilaterally and without turning to the forums provided by international organizations. The administration of use rules for common-property resources, by contrast, is apt to require the establishment of organizations, though the results may be more equitable than those arising from shifts in jurisdictional boundaries.[50]

Nonetheless, policy instruments suitable for use by organizations are by no means absent from the realm of international regimes. The International Whaling Commission has the authority to adjust annual harvest quotas for individual species of great whales. The International Monetary Fund can lay down specific conditions in granting loans to countries experiencing balance-of-payments problems. The International Coffee Agreement allows for the allocation of export shares among its members. And the proposed International Seabed Authority would be able to make use of a relatively complex system of licenses to regulate the production of manganese nodules from the deep seabed. The ability of these organizations to operate autonomously in using such instruments

48. For further discussion see Giandomenico Majone, "Choice among Policy Instruments for Pollution Control," *Policy Analysis* 2 (1976), 589–613.

49. Changes in regimes for marine fisheries arising from unilateral extensions of jurisdiction on the part of coastal states exemplify this prospect. In the case of the United States, the transition was accomplished through the passage of the Fishery Conservation and Management Act of 1976. For further discussion see Oran R. Young, "The Political Economy of Fish: The Fishery Conservation and Management Act of 1976," *Ocean Development and International Law* 10 (1982), 199–273.

50. For a case in point see the discussion of the regime for deep-seabed mining in chapter 5 below.

may be severely limited. Compliance can also become a major problem in the use of some of these instruments (for example, export quotas for coffee).[51] But these facts do not suggest a qualitative difference between the use of policy instruments in international regimes and their use in domestic regimes.[52] In short, though the use of these instruments is limited by the characteristic weaknesses of organizations associated with international regimes, policy instruments are certainly not irrelevant at the international level.

CONCLUSION

Though there are great variations among international regimes, they are all social institutions. This suggests, among other things, that regimes are products of human interactions; specific regimes are always created rather than discovered. Formalization is not a necessary condition for the existence or operation of any given regime. This fact undoubtedly accounts for certain methodological problems that arise in efforts to study international regimes systematically. But the recognition of this attribute is a prerequisite for the achievement of analytic success.

In closing this chapter, let me lay out an agenda of questions that require consideration in the analysis of any international regime:

1. *Institutional character*. What are the principal rights, rules, and social choice procedures of the regime? How do they structure the behavior of individual actors to produce a stream of collective outcomes?
2. *Jurisdictional boundaries*. What is the coverage of the regime in terms of functional scope, areal domain, and membership? Is this coverage appropriate under the prevailing conditions?
3. *Conditions for operation*. What conditions are necessary for the regime to work at all? Under what conditions will the operation of the regime yield particularly desirable results (for example, economic efficiency, distributive justice, ecological balance)?
4. *Consequences of operation*. What sorts of outcomes (either individual or collective) can the regime be expected to produce? What are the appropriate criteria for evaluating these outcomes?
5. *Regime dynamics*. How did the regime come into existence, and what is the

51. Bart S. Fisher, "Enforcing Export Quota Agreements: The Case of Coffee," *Harvard International Law Journal* 12 (1971), 401–435.

52. It is not necessary to subscribe to Marxian precepts to realize that domestic as well as international regimes are sometimes influenced greatly by pressures from actors who are, in principle, subject to regulation under the terms of these regimes. This is, in fact, the central insight of the "capture" theory of regulation.

likelihood that it will experience major changes in the foreseeable future? Does the regime include transformation rules that are likely to be effective?

Adherence to such an agenda should produce a growing body of comparative studies dealing with international regimes. Over time, these studies will improve our ability to construct powerful generalizations about this fundamental, yet still poorly understood, international phenomenon.

CHAPTER TWO

Patterns of International Cooperation:
Institutions and Organizations

Students of international relations, like most other social scientists, regularly use the terms *institution* and *organization* interchangeably. Yet this practice gives rise to serious problems. Not only does it sow confusion by conflating two distinct phenomena, but it also forecloses any possibility of analyzing the complex and interesting relationships between institutions and organizations.[1] This chapter sets out to remedy this situation, with particular reference to the study of international regimes.[2] It begins with a discussion of the distinction between international institutions and international organizations. It then explores systematically the linkages between institutions and organizations, analyzing roles organizations play in the administration and maintenance of institutions as well as opportunities for organizations to contribute to regime formation or the creation of institutions. In keeping with the general plan of this book, the chapter makes extensive use of examples pertaining to international resource and environmental regimes.[3] But the basic arguments are generic; they apply to the entire domain of international relations.

1. For other statements regarding the importance of distinguishing between institutions and organizations see John Gerard Ruggie, "International Responses to Technology: Concepts and Trends," *International Organization* 29 (1975), 557–583, and Friedrich Kratochwil and John Gerard Ruggie, "International Organization: A State of the Art or an Art of the State," *International Organization* 40 (1986), 753–775.

2. To obtain a clear sense of the state of the regimes literature consult Stephen D. Krasner, ed., *International Regimes* (Ithaca: Cornell University Press, 1983); Oran R. Young, "International Regimes: Toward a New Theory of Institutions," *World Politics* 34 (1986), 104–122; and Robert O. Keohane, "The Study of International Regimes and the Classical Tradition in International Relations," unpublished paper, Cambridge, Mass., June 1986.

3. A helpful reference work in this area is Lynton Keith Caldwell, *International Environmental Policy: Emergence and Dimensions* (Durham: Duke University Press, 1984).

Institutions and Organizations

Institutions are social practices consisting of easily recognized roles coupled with clusters of rules or conventions governing relations among the occupants of these roles. The rules that link institutionalized roles and, therefore, form the superstructure of institutions ordinarily encompass sets of rights or entitlements (for example, voting rights or property rights) as well as sets of behavioral prescriptions (for example, rules governing eligibility to vote or the transfer of property from one owner to another). Electoral systems are institutions in which candidates for office interact with voters according to explicit rules specifying when elections will take place, who may become a candidate, who is eligible to vote, and what it takes to win election.[4] Markets are institutions in which buyers and sellers interact with one another on the basis of well-defined rules pertaining to various aspects of exchange relationships, such as the nature of contracts, permissible forms of advertising, and liability arrangements.[5] Similarly, structures of property rights are institutions in which owners and nonowners interact in terms of clear-cut rights (for example, the use and exclusion rights of owners) and rules (for example, zoning restrictions, prohibitions on the use of property to harm others, and provisions relating to trespass).[6] In every case, the existence of an institution sets up a network or pattern of behavioral relationships that lends order or predictability to human affairs.[7]

Organizations, by contrast, are material entities possessing physical locations (or seats), offices, personnel, equipment, and budgets. Equally important, organizations generally possess legal personality in the sense that they are authorized to enter into contracts, own property, sue and be sued, and so forth. In political terms, we often focus on governmental agencies, bureaux, or commissions in thinking about organizations. Thus, the Bureau of Land Management, the Forest Service, the National Park Service, the North Pacific Fisheries Management Council, and the Alaska Commercial Fisheries Entry Commission are all organizations. But there are of course numerous organizations in the private sector as well: the American Medical Association, the American Bar

4. Duncan Black, *The Theory of Committees and Elections* (Cambridge: Cambridge University Press, 1958), and Douglas W. Rae, *The Political Consequences of Electoral Laws* (New Haven: Yale University Press, 1971).

5. Robert Haveman and Kenyon A. Knopf, *The Market System,* 3d ed. (Santa Barbara: Wiley, 1978).

6. A. Irving Hallowell, "The Nature and Function of Property as a Social Institution," *Journal of Legal and Political Sociology* 1 (1943), 115–138.

7. Note that this tells us nothing about the degree to which the consequences that institutions produce are just, fair, or otherwise desirable.

Association, the National Petroleum Council, the National Rifle Association, the National Wildlife Federation.

International society is well-stocked both with institutions and with organizations. The states system itself is an elaborate institution governing not only interactions among states but also the activities of individuals and other juridical persons (for example, corporations) who are treated as nationals of states under this system.[8] So is the emerging public order of the oceans articulated initially in the Geneva Conventions of 1958 and recast in the Law of the Sea Convention of 1982.[9] Similarly, the international monetary system is an institution, albeit one that is nested within the broader international economic order and that has been undergoing dramatic changes in recent decades.[10] Even war is a social institution, a fact that undoubtedly contributes to the difficulty of eliminating organized violence from human affairs.[11]

The United Nations, on the other hand, is an organization. So also are the North Atlantic Treaty Organization, the Organization of American States, and the Organization of Petroleum Exporting Countries. Similarly, there are now several thousand international nongovernmental organizations, ranging from commercial agencies such as the International Air Transport Association through humanitarian organizations such as Amnesty International, religious organizations such as the World Council of Churches, environmental organizations such as the World Wildlife Fund, scientific organizations such as the International Council of Scientific Unions, and organizations representing the concerns of indigenous peoples such as the World Council of Indigenous Peoples.[12]

To sharpen and enrich this distinction between institutions and organizations, I shall describe a few illustrative cases in greater detail.

8. Hedley Bull, *The Anarchical Society: A Study of Order in World Politics* (New York: Columbia University Press, 1977). For an account of the operation of the states system with respect to natural resources see Richard B. Bilder, "International Law and Natural Resources Policies," *Natural Resources Journal* 20 (1980), 451–486.

9. For an institutional perspective on the public order of the oceans consult Myres S. McDougal and William T. Burke, *The Public Order of the Oceans* (New Haven: Yale University Press, 1962).

10. Robert O. Keohane and Joseph S. Nye, *Power and Interdependence: World Politics in Transition* (Boston: Little, Brown, 1977).

11. Hedley Bull, "War and International Order," in Alan James, ed., *The Bases of International Order* (Oxford: Oxford University Press, 1973), 116–132; Quincy Wright, *A Study of War* (Chicago: University of Chicago Press, 1942); and Michael Howard, *War in European History* (Oxford: Oxford University Press, 1976).

12. For a more general account of the role of organizations in international society see Harold K. Jacobson, *Networks of Interdependence: International Organizations and the Global Political System* (New York: Knopf, 1979).

The regime for the Svalbard Archipelago was established under the terms of the treaty of 9 February 1920 relating to Spitsbergen.[13] The archipelago is a collection of islands located some 600 miles northwest of the north coast of Norway and covering 62,400 square kilometers (about the size of Belgium and the Netherlands combined). The treaty, which now has forty signatories including both the United States and the Soviet Union, originated in the larger settlement of issues at the end of World War I. In essence, the regime articulated in the treaty of 1920, which entered into force in 1925, couples a recognition of Norwegian sovereignty over the archipelago with a series of commitments on the part of Norway to respect all previously established rights in the area, to allow nationals of all the signatories equal access to exploit the natural resources of the archipelago, and to maintain the archipelago in a demilitarized state. Legally, therefore, the Svalbard Archipelago is a part of Norway. But the Norwegian government does not have the authority to exclude others from using its resources including both minerals and fish, and Norway has an international obligation to prevent the use of it for warlike purposes.

The Spitsbergen treaty does not establish a specialized organization to administer the Svalbard regime. Rather, the government of Norway handles all administrative functions for the area. As Østreng says, "Svalbard is administered by a Governor, who combines the functions of district judge and revenue officer."[14] Similarly, "the Commissioner of Mines is responsible for ensuring the observance of the Mining Code, which applies to all nationalities in Svalbard."[15] There have been disagreements from time to time regarding the application of Norwegian regulations to the Russian mining operations located at Barentsburg and Pyramiden as well as the rules that would govern efforts to extract natural resources from the continental shelves surrounding Svalbard. But no one has proposed the creation of a separate organization for the administration of the regime.

As a second example, consider the regime for renewable resources set forth in the provisions of the convention of 11 September 1980 on the Conservation of Antarctic Marine Living Resources (the Southern Ocean convention).[16] This convention, intended to complement the Antarctic Treaty of 1959, now includes sixteen Contracting Parties and

13. For a general account see Willy Østreng, *Politics in High Latitudes: The Svalbard Archipelago* (London: C. Hurst, 1977).
14. Ibid., 3.
15. Ibid.
16. James N. Barnes, "The Emerging Antarctic Living Resources Convention," *Proceedings* of the American Society of International Law, 1979, 272–292.

covers the marine environment located between Antarctica proper and the Antarctic Convergence. It lays out a restricted common-property regime designed to ensure conservation (defined to include rational use) of marine living resources. The Contracting Parties agree to engage in joint decision making on the harvest of marine life as well as the promulgation of regulations governing harvesting activities (for example, the designation of protected species, size limits, open and closed seasons, and gear restrictions). Additionally, the parties commit themselves to taking an ecosystems approach in that they agree to maintain ecological relationships between harvested and other populations and to prevent changes in the marine ecosystems of the area that are not reversible over two or three decades.

To administer this regime, the convention, which entered into force in 1982, establishes an organizational structure composed of a Commission for the Conservation of Antarctic Marine Living Resources (Articles 7–13), a Scientific Committee for the Conservation of Antarctic Marine Living Resources (Articles 14–17), and a secretariat headed by an executive secretary (Articles 18–19). The seat of the organization is at Hobart, Tasmania. Article 9 grants the commission extensive authority to make decisions to fulfill the objectives of the regime, and Articles 17 and 19 set forth specific provisions for personnel and a budget. The convention states explicitly (Article 8) that the commission will have both legal personality and the "legal capacity necessary to perform its functions and achieve the purposes of this Convention." Additionally, Article 23 calls on the organization created under the terms of the convention to cooperate with a number of other organizations, including the Antarctic Treaty Consultative Meetings, the United Nations Food and Agriculture Organization, and such other intergovernmental and nongovernmental organizations as may prove helpful to its operations.

Another pattern emerges from a consideration of the regime for deep-seabed mining articulated in Part XI of the 1982 Convention on the Law of the Sea.[17] This regime rests on the principle that the deep seabed (called the Area) and its resources are the common heritage of mankind (Article 136). No state is permitted to claim sovereignty or to assert exclusive management authority over the Area. Nor is any state or other juridical person (for example, a corporation) allowed to assert private-property rights in the deep seabed or its resources. Instead, the convention establishes a common-property regime. But participants

17. For relevant background see the papers in parts 3, 4, and 5 of Bernard H. Oxman, David D. Caron, and Charles L. O. Buderi, eds., *Law of the Sea: U.S. Policy Dilemma* (San Francisco: ICS Press, 1983).

cannot use the deep seabed on an open-to-entry basis as in the case of the traditional common-property regime for the high seas. On the contrary, they must abide by a highly restricted common-property regime entailing a complex framework of rules governing the terms of entry, the allocation of tracts, environmental regulations, technology transfer, and the distribution of benefits from mining (Sections 2 and 3 of Part XI).

The convention, which has not yet entered into force, provides for a comparatively elaborate organizational structure to administer the regime. Known collectively as the International Seabed Authority, the organization is made up of an assembly, a council, a secretariat, and an Enterprise, which may engage in actual mining operations under certain conditions. The seat of the Authority will be Kingston, Jamaica. The Authority is charged with responsibility not only for administering a complex of rules governing the mining activities of member states and corporations but also for overseeing the provisions of the regime regarding technology transfer and the distribution of benefits from deep-seabed mining. It may even function as an operating authority in its own right by activating the provisions of the regime pertaining to the Enterprise. The Authority will have the power to raise money from mining carried out under its jurisdiction as well as to borrow funds to carry out its operations. The Authority will have international legal personality; it will also "enjoy immunity from legal process except to the extent that the Authority expressly waives this immunity in a particular case" (Article 178). Given the character of the seabed regime as a highly restricted common-property arrangement, the performance of the Authority will be critical to the operation of the entire system.

With the distinction between institutions and organizations clearly established, we are ready to turn to an analysis of the relationships between the two. Some international regimes require organizations to administer them; others do not. Whereas many international organizations are explicitly linked to regimes, others are freestanding entities functioning apart from, or in the absence of, operative regimes. To think about these relationships in an orderly fashion, it will help to construct a simple 2 × 2 matrix. As Figure 1 indicates, the matrix identifies four distinct cases. Where regimes or institutional arrangements operate in the absence of organizations, I shall describe the result as international anarchy. The addition of organizations transforms international anarchy into a condition of civil society. In cases where organizations function independently of regimes, I shall speak of freestanding organizations. Finally, the absence of both institutions or regimes and organizations gives rise to a situation that I shall characterize as an international state of nature.

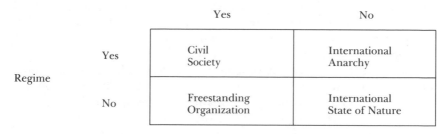

Figure 1. Matrix of institutions and organizations

Under what conditions can regimes perform satisfactorily in the absence of organizations? Are there cases in which the formation of a regime is sufficient to ensure that some organization will come into existence? Do organizations assume fundamentally different roles in connection with various types of international regime? Can freestanding organizations play a leading part in regime formation at the international level? Does a true state of nature prevail in many sectors of international society? What are the consequences when it does? These are the questions that any analysis of the linkages between international institutions or regimes and international organizations must address.

INTERNATIONAL ANARCHY

Anarchy, as its proponents regularly point out, is not a condition of disorder or chaos. Rather, it is a social state featuring institutions or regimes that operate in the absence of central organizations to administer them.[18]

Institutional arrangements that eliminate the need for organizations altogether or at least minimize the role of organizations have real attractions. As philosophically inclined analysts often observe, the absence of administrative agencies serves to protect the freedom of those participating in social practices.[19] To be sure, it is erroneous to suppose that individual participants will find themselves at liberty to do just as they please in the absence of organizations or administrative agencies. We are all familiar with the effects of socialization and social pressure as factors militating against deviation from the dictates of social conventions. What is more, the absence of organizations may protect negative free-

18. Daniel Guerin, *Anarchism* (New York: Monthly Review Press, 1970).
19. Robert Nozick, *Anarchy, State, and Utopia* (New York: Basic Books, 1974).

dom at some cost in terms of positive freedom.[20] Nonetheless, it is indisputable that eliminating organizations will protect individual participants from interference in their affairs on the part of central authorities seeking to impose some conception of the general will (like that derived from the ideas of Rousseau and his intellectual heirs).

Additionally, anarchical arrangements facilitate efforts to sidestep a range of problems that analysts now group under the rubric of nonmarket failures.[21] Of course, organizations are costly to operate. In domestic societies, for instance, governments often control the expenditure of 25–50 percent of the gross national product, a substantial portion of which goes to cover the costs of administration in contrast to substantive programs. Moreover, organizations frequently perform inefficiently, wasting resources and engaging in redundant activities. They typically give rise to the rigidities and stultifying effects associated with bureaucratic behavior. The activities of organizations commonly generate unforeseen and unintended consequences that actually impede progress toward social goals. As a result, it is no longer acceptable to justify organizations on the grounds that they may contribute to the achievement of social goals and that they are likely to be harmless in any case. On the contrary, it is appropriate to apply a benefit/cost test to the performance of organizations and to resist the introduction of organizations that fail to pass such a test.

Students of international relations generally describe international society as anarchical.[22] By contrast with domestic society, both the extent and the authority of central organizations are sharply limited at the international level. During much of the postwar era, commentators pointed to this condition as a fundamental defect in international society. Many called for world government to eliminate war and to solve other international problems.[23] Even those of a more pragmatic bent argued the case for a dramatic expansion in the role of specific international organizations. The United Nations, for example, arose in a social environment dominated by such thinking. So did a wide variety of functionally specific organizations, like the International Monetary Fund and the International Whaling Commission. Recently, however, a pro-

---

20. On the distinction between negative freedom or "freedom from" and positive freedom or "freedom to," see Isaiah Berlin, *Four Essays on Liberty* (Oxford: Oxford University Press, 1969).

21. See Charles Wolf, Jr., "A Theory of Nonmarket Failure: Framework for Implementation Analysis," *Journal of Law and Economics* 22 (1979), 107–139.

22. Bull, *The Anarchical Society*.

23. For a well-known example see Grenville Clark and Louis B. Sohn, *World Peace through World Law* (Cambridge: Harvard University Press, 1958).

nounced mood of skepticism has set in regarding these prescriptions. As the catalog of nonmarket failures grew, analysts began to express persistent doubts about the social value of organizations (or governments) at many social levels. As a result, we are now witnessing a growth of interest in international society as a source of relevant experience with anarchical arrangements or regimes involving the minimum in the way of administrative apparatus.[24]

Zonal arrangements constitute the preeminent category of anarchical institutions in international society.[25] These arrangements feature geographically or areally delimited zones within which individual members (usually states) exercise jurisdiction or exclusive management authority. There is no need for central administrative agencies in connection with zonal arrangements. In their zones, individual states make and enforce their own rules without consulting others or seeking their consent. Outsiders must conform to these rules or simply refrain from operating in the zones of others. Some recent developments suggest an erosion of the dominance of zonal arrangements in international society. Most states now acknowledge international obligations regarding matters such as the maintenance of human rights, the protection of wildlife, and the control of pollution. Similarly, influential states often seek to bring pressure to bear on others to take certain actions within their own jurisdictional domains (for example, to put an end to whaling operations or curb the trade in endangered species). Yet the enclosure movement that has arisen in connection with marine areas during the postwar era makes it clear that zonal arrangements remain highly attractive in international society. Coastal states now operate fishery-conservation zones, pollution-control zones, and arrangements featuring exclusive management authority over the outer continental shelves.[26] With the advent of exclusive economic zones covering extensive marine areas, moreover, zonal arrangements have gained a greatly enhanced role in international society.[27]

A second class of anarchical arrangements in international society encompasses exchange systems. The essential element of such systems is

24. For a thoughtful account of the supply of international public goods in the absence of central organizations see Charles P. Kindleberger, "International Public Goods without International Government," *American Economic Review* 76 (1986), 1–13.

25. For a discussion of zonal arrangements with particular reference to natural resources see Oran R. Young, *Resource Management at the International Level* (London and New York: Pinter and Nichols, 1977), especially chap. 5.

26. For an account that looks upon the enclosure movement with favor see Ross D. Eckert, *The Enclosure of Ocean Resources* (Stanford: Hoover Institution Press, 1979).

27. On the events culminating in the advent of EEZs consult Ann L. Hollick, *U.S. Foreign Policy and the Law of the Sea* (Princeton: Princeton University Press, 1981).

the assignment of transferable property rights in valued goods to individual members of international society together with a reliance on the operation of market mechanisms to organize interactions among the holders of these rights. So long as the resultant markets do not suffer from severe market failures, there is no need to set up extensive organizations to administer institutional arrangements of this sort. Of course, exchange systems are widespread with respect to conventional private goods. In fact, the role of such systems has tended to expand in international society during the course of the twentieth century as a consequence of the reduction of trade barriers and the growth of competitive markets at the international level.[28] Today, there is serious interest in expanding the role of such systems even further by creating property rights or limited property interests in additional goods. Many commentators have suggested the establishment of property rights in the manganese nodules of the deep seabed,[29] and there are those who espouse the creation of transferable property rights in broadcast frequencies as a method of managing the use of the electromagnetic spectrum.[30]

A third class of anarchical arrangements features decentralized coordination. Under such arrangements, the members of international society agree to common rules governing well-defined activities with the understanding that individual members will take responsibility for implementing these rules within their own jurisdictions or administrative zones. The arrangements established under the provisions of the 1971 Convention on Wetlands of International Importance Especially as Waterfowl Habitat (the Ramsar convention) and the 1973 Agreement on the Conservation of Polar Bears exemplify this option. With the advent of exclusive economic zones for marine areas, much the same can be said of the pollution-control regimes for the Mediterranean Sea, the Persian Gulf, or the Caribbean area established under the auspices of the Regional Seas Programme of the United Nations Environment Programme (UNEP).[31] Going a step further, the members of international society sometimes set up self-help arrangements that do not require central organizations. Such practices involve mutually agreed upon rules for activities that cut across zonal boundaries (for example, com-

---

28. For relevant background see Charles Lipson, "The Transformation of Trade: The Sources and Effects of Regime Change," in Krasner, *International Regimes*, 233–271.

29. Eckert, *Enclosure of Ocean Resources*, chap. 8.

30. For further discussion of such ideas consult Francis T. Christy, Jr., "Property Rights in the World Ocean," *Natural Resources Journal* 15 (1975), 695–712, and Seyom Brown, Nina W. Cornell, Larry L. Fabian, and Edith Brown Weiss, *Regimes for the Ocean, Outer Space, and Weather* (Washington, D.C.: Brookings, 1977).

31. Caldwell, *International Environmental Policy*, chap. 5.

mercial shipping) or take place in some international commons (for example, the oceans, the atmosphere, outer space, Antarctica) in contrast to areas under the jurisdiction of individual states. But they feature self-help systems in that individual members assume responsibility for implementing the mutually agreed upon rules (usually with respect to their own nationals). The traditional regime for the high seas exemplifies this option. Emerging arrangements regarding the use of outer space follow a similar pattern. The regime for Antarctica, established under the Antarctic Treaty of 1959, offers an interesting variant on this pattern because it permits each of the Consultative Parties to monitor the activities of the others in Antarctica to ensure that these activities are in compliance with the provisions of the regime.[32]

A final method of avoiding central organizations is to agree on common rules for an area or activity but then to delegate authority for the administration of the regime to an individual member or even to an outside entity. The regime for Svalbard, which relies on Norway for administration, is a case in point. Even more intriguing arrangements are now in place in the cases of the regimes set up under the 1973 Convention on International Trade in Endangered Species of Wild Fauna and Flora (CITES) and the 1979 Convention on the Conservation of Migratory Species of Wild Animals (the Bonn convention).[33] These conventions delegate responsibility for the administration of the regimes to UNEP or, even more specifically, to the executive director of UNEP. In the case of CITES, the executive director of UNEP has gone a step further and passed administrative responsibility on to the International Union for the Conservation of Nature and Natural Resources (IUCN) and the World Wildlife Fund.[34] The result is a full-fledged international regime administered by nongovernmental organizations that are freestanding and that engage in numerous other activities in addition to their efforts on behalf of the CITES regime.

Of course, there are sharp limitations on what can be achieved through the operation of anarchical institutions. Above all, anarchy necessitates the selection of certain types of regimes over others. It is easy enough, for example, to set up an open-to-entry common-property regime on an anarchical basis. But restricted common-property arrangements, like the whaling regime or the deep-seabed mining regime,

---

32. Philip W. Quigg, *A Pole Apart: The Emerging Issue of Antarctica* (New York: McGraw-Hill, 1983).
33. Simon Lyster, *International Wildlife Law* (Cambridge: Grotius Publications), chaps. 12 and 13.
34. Caldwell, *International Environmental Policy*, 189–192.

typically require organizations to set quotas, decide on moratoriums, promulgate regulations, control production levels, and so forth. Similarly, regimes that encompass provisions for redistributing wealth (for example, the deep-seabed mining regime) will generally require organizations to collect revenues and to administer redistributive programs.[35] Nonetheless, critics frequently exaggerate the limitations of anarchical institutions. Ingenuity is sometimes sufficient to solve collective-action problems without resorting to the creation of organizations or administrative agencies. A brief account of some of the putative limits on anarchy in international society will serve to clarify this proposition.

A wide spectrum of international regimes rely on central organizations to collect and analyze pertinent information. The fur-seal regime, the whaling regime, the regime for Antarctic living resources, and the international seabed regime, for example, all feature committees or commissions charged with data collection and analysis. But a little reflection will suffice to demonstrate that alternative arrangements can work perfectly well in specific cases. Freestanding organizations like the International Bureau for Whaling Statistics (located in Sandefjord, Norway) in connection with commercial whaling or the IUCN in connection with endangered species[36] play a key role in data collection and analysis for some social practices. Private organizations like the Stockholm International Peace Research Institute and the International Institute of Strategic Studies have come to play significant roles as sources of information pertaining to arms-control regimes. In some cases, the governments of individual members can do a better job of data collection and analysis than international organizations can realistically hope to do. It is hard to imagine the governments of members states relying on international organizations, for example, to verify compliance with the terms of arms-control arrangements, such as those established under the Partial Nuclear Test Ban Treaty of 1963 or the Nuclear Nonproliferation Treaty of 1968.[37]

Collective decision making is another functional requirement that often serves to justify the transition from anarchy to civil society. Yet it is perfectly possible to arrive at collective choices in the absence of central

35. For a more general discussion of principles of equitable distribution of the world's wealth and natural resources see Oscar Schachter, *Sharing the World's Resources* (New York: Columbia University Press, 1977).

36. The *Red Data Book*, published and periodically updated by the IUCN, has acquired considerable influence in international efforts to protect endangered species.

37. Abram Chayes, "An Enquiry into the Workings of Arms Control Agreements," *Harvard Law Review* 85 (1972), 905–969.

authorities. Of course, exchange systems based on relatively unrestricted structures of private-property rights exemplify this proposition. But open-to-entry common-property systems in which allocation is accomplished through the operation of the principle of first come, first served (like the traditional regime for the marine fisheries) also yield collective choices in the absence of organizations. The point to stress in this connection, therefore, is that the need to establish organizations is a function of the type of institutional arrangements created rather than of the need for collective decision making per se. The regime for deepseabed mining articulated in the provisions of the 1982 Law of the Sea Convention will clearly require an organization resembling the proposed International Seabed Authority. Still, it would have been possible to create a regime for deep-seabed mining based on transferable property rights requiring no administrative apparatus beyond some minimal international registry for mining claims.[38]

Many observers treat compliance as a critical problem requiring central organization. In fact, they attribute numerous international problems to the absence of central authorities capable of monitoring compliance and enforcing the rules of institutional arrangements. But it is easy to exaggerate this limitation of anarchy as well.[39] No institution can operate effectively unless the great majority of those subject to its rights and rules comply with them voluntarily (or at least in the absence of overt coercion) most of the time. The fact that international society is well-stocked with viable regimes testifies to the pervasiveness of voluntary compliance, even at the international level. Beyond this, however, decentralized compliance mechanisms are often effective in international society. It is possible to bring pressure to bear on some violators (for example, those who fail to comply with international pollution standards) in the municipal courts of member states. Individual members can threaten to impose sanctions on others who violate the provisions of international regimes. This is what the United States has done in the cases of the whaling moratorium and the trade in elephant ivory. In addition, the prospect that others will withdraw from a regime, initating

38. Eckert, *Enclosure of Ocean Resources*, chap. 8.

39. For general discussions of compliance at the international level consult Oran R. Young, *Compliance and Public Authority: A Theory with International Applications* (Baltimore: John Hopkins University Press, 1979), and Roger Fisher, *Improving Compliance with International Law* (Charlottesville: University Press of Virginia, 1981). For an account that focuses on resource regimes see William T. Burke, Richard Legatski, and William W. Woodhead, *National and International Law Enforcement in the Ocean* (Seattle: University of Washington Press, 1975).

a process in which a complex social practice unravels, is often sufficient to deter individual members from violating the terms of international regimes. It seems clear, for example, that the danger of touching off destructive trade wars is a major consideration in the behavior of individual states regarding compliance with the terms of the international-trade regime.

Somewhat similar comments are in order with respect to the problem of dispute resolution. While it is convenient to be able to turn to a well-established organization to handle conflicts among participants in a social practice, it is hardly a necessity. Disputants can always resolve their differences through direct negotiation, with or without the assistance of third parties. They can resort to ad hoc arbitral tribunals established for the purpose of settling specific disputes (for example, the Bering Sea arbitration regarding fur seals in the 1890s). Or they can turn to established organizations, like the International Court of Justice, that are freestanding in the sense that they are independent of the institutional arrangement in question. The role of the Court in resolving the North Sea maritime boundary issue in the 1960s and the Gulf of Maine controversy between Canada and the United States in the 1980s exemplifies this option. What is more, organizations set up to resolve disputes arising from the operation of specific regimes are frequently ignored. Those desiring to settle their differences can generally find a way to do so, whereas those unwilling to settle are not likely to accept the jurisdiction of a court or specialized tribunal, such as the proposed Seabed Chamber of the new regime for deep-seabed mining.

Finally, critics of anarchical arrangements frequently point out that such practices are slow to adjust to rapidly changing circumstances. But it is far from clear that the introduction of central organizations will solve this problem. Powerful interest groups that benefit from existing institutional arrangements often dominate the actions of organizations. Additionally, organizations are notorious for creating bureaucratic procedures that impede change regardless of shifting circumstances in the external environment. Organizations may actually become a hindrance to adjustment rather than a vehicle for adapting institutional arrangements to changing circumstances. This observation suggests that there is much to be said for arrangements located in the transition zone between anarchy and civil society, such as the regime for Antarctica under which the full members hold biennial Consultative Meetings or the whaling regime under which the signatories hold an annual meeting. These arrangements allow for responses to changing circumstances without giving rise to the rigidities of bureaucratized organizations.

CIVIL SOCIETY

Despite the prominent place of anarchical arrangements in international society, many international regimes do encompass central organizations. The Commission for the Rhine River, established in the early nineteenth century, administers the evolving international regime for the Rhine River Basin. The International Whaling Commission has come to play an increasingly significant role both in setting quotas and in making decisions on moratoriums under the terms of the international regime for whaling. The International Pacific Halibut Commission, set up to administer the North Pacific halibut regime, has acquired considerable influence because it has its own staff and research capability. The Helsinki Commission administers the regime for the Baltic created under the provisions of the Convention on the Protection of the Marine Environment of the Baltic Sea Area of 1974. And it would be easy to augment this list with numerous related examples. If anything, the role of organizations linked to institutional arrangements at the international level has grown during the recent past.

What accounts for the importance of organizations even in the anarchical environment of international society? All those who analyze negotiations regarding the establishment of international regimes will be familiar with the issue of whether to fold organizations into institutional arrangements as an element in the bargaining process. In some instances, incentives to establish central organizations spring from the substantive nature of the problems under consideration. For instance, some observers reason that there will be a need for organizations to administer regimes dealing with international commons (for example, oceans or outer space) or transboundary problems (for example, flow resources such as water or migratory resources such as fish or birds).[40] Yet it is easy to make too much of this line of reasoning. The traditional regime for the high seas was a self-help arrangement under which individual states applied common rules to their own nationals. The current regime for Antarctica requires remarkably little in the way of central organizations. Most of the migratory-bird regimes establish systems of decentralized coordination under which individual parties are

40. For a range of perspectives on such matters consult Per Magnus Wijkman, "Managing the Global Commons," *International Organization* 36 (1982), 511–536; Marvin S. Soroos, "The Commons in the Sky: The Radio Spectrum and Geosynchronous Orbit as Issues in Global Policy," *International Organization* 36 (1982), 665–677; and the special issue on "Managing International Commons," of *Journal of International Affairs* 31 (Spring-Summer 1977).

responsible for implementing common rules within their own jurisdictions.

Still, there are problems that do call for central organizations. When the basic objective is to devise an institutionalized procedure for reaching agreement on the designation of sites for inclusion in protected categories, for example, it is difficult to avoid the creation of a commission or committee capable of reviewing proposals and making choices. Consider in this connection the role of the World Heritage Committee set up under the terms of the 1972 Convention concerning the Protection of the World Cultural and Natural Heritage (the World Heritage Convention) to select sites for inclusion on a World Heritage List.[41] Similarly, when the problem revolves around the development of cost-sharing arrangements for the provision of some public good, such as peacekeeping or coordinated research, the case for central organizations will be strong.[42] In such instances, it is hard to devise any alternative to central organizations to collect contributions and to apply them to the supply of the relevant public goods.[43]

A more compelling argument centers on the proposition that incentives to create organizations in conjunction with international regimes flow from the character of the regimes under consideration. If the participants elect to create an open-to-entry common-property regime, for example, they will not experience strong incentives to establish a central organization. Much the same is true of regimes featuring relatively unrestricted private-property systems coupled with competitive markets. Restricted common-property systems, by contrast, almost always require some sort of organization. At a minimum, there will be a need for an organization to establish overall limits on efforts or certain types of behavior (for example, total allowable catches in fisheries, catch quotas for whales, or effluent standards for pollution regimes). Organizations become even more important in conjunction with regimes intended to control entry or regulate the activities of individual resource users. Consider the role of the Commission for the Conservation of Antarctic Marine Living Resources in this light. When regimes include additional restrictions, such as the transfer of technology under complex rules, the regulation of production levels to control markets, the re-

41. See Lyster, *International Wildlife Law*, chap. 11.
42. Consider the emerging role of the United Nations University as an organization dedicated to promoting the public good of coordinated research.
43. Even here, situations fulfilling the conditions of Olsonian privileged groups can constitute exceptions. See Mancur Olson, Jr., *The Logic of Collective Action* (Cambridge: Harvard University Press, 1965).

distribution of benefits derived from the exploitation of resources, or the creation of international operating authorities to exploit resources directly, the need for central organizations naturally grows. This is why the organizational structure articulated under the provisions of the regime for the deep seabed is so extensive and elaborate. As I have already suggested, however, there is nothing inevitable about these arrangements. It would have been possible to set up a regime for the deep seabed requiring no central organization, other than some simple registry for the filing of claims. The need for the elaborate organizational structure incorporated into the provisions for the International Seabed Authority is a function of the character of the regime articulated in the bargaining process in the Law of the Sea Conference.

There are also circumstances in which incentives to set up organizations in conjunction with international regimes stem from the dynamics of the bargaining process itself. Consider the problem of data collection and analysis. Most international regimes require the collection and analysis of data to support decisions on the use of resources (for example, the designation of endangered species), to monitor the compliance of participants, and so forth. As I argued in the preceding section, however, there are often ways to solve such problems that do not require the creation of central organizations. Yet it is easy to understand how the establishment of central organizations to handle the tasks of data collection and analysis can acquire the character of a salient or prominent solution from the perspective of those bargaining over the terms of an international regime.[44] Such organizations possess an aura of impartiality and credibility, a factor of considerable importance in the competitive/cooperative atmosphere of a bargaining process. What is more, organizations of this type can fill the need for tangible symbols of the success of efforts to create negotiated regimes without placing severe restrictions on the freedom of action of the participants. There is nothing unusual, therefore, about situations in which the establishment of organizations in conjunction with institutional arrangements owes more to the character of the bargaining process itself than to the character of the regimes actually created.

Finally, it is worth noting the role of intellectual fashions as determinants of the choices of those striving to create international regimes. Until recently, for example, there was a distinct tendency to focus on

44. On the concept of salient or prominent solutions in connection with bargaining processes see Thomas C. Schelling, *The Strategy of Conflict* (Cambridge: Harvard University Press, 1960).

market failures and to envision roles for organizations in correcting for such failures.[45] Most of the organizations (generally known as councils) established under the terms of the international regimes for commodities, such as coffee, sugar, rubber, and tin, exemplify this perspective. In a sense, the negotiation of the terms of the proposed International Seabed Authority during the 1970s and early 1980s represented a kind of culmination of this trend. Today, by contrast, it has become fashionable to focus on nonmarket failures, which point to a wide range of problems afflicting governments at the domestic level and organizations, such as the United Nations or numerous functionally specific agencies, at the international level.[46] The implication of this line of reasoning is that we should be cautious about the creation of new organizations, relying as much as possible on institutional arrangements capable of coping with collective-action problems in the absence of organizations. Although both market failures and nonmarket failures are real enough, there is no clear-cut procedure for calculating the relative weight of these problems in specific situations. As a result, those negotiating the terms of specific institutional arrangements are left either to fall back on ideological preferences or to respond to the winds of intellectual fashion regarding the relative merits of establishing central organizations in creating institutional arrangements. This observation suggests that we should not be surprised by a rising tide of antipathy to central organizations in conjunction with international regimes. The problems afflicting current efforts to activate the proposed International Seabed Authority undoubtedly reflect, at least in part, this development. But it would be a mistake to suppose that this antipathy will last indefinitely. On the contrary, it is safe to anticipate that there will be a resurgence of interest in roles for central organizations in due course.[47]

By contrast, dominant actors (or hegemons) seeking to forge the links of imposed regimes may seem, at first, to have little incentive to establish central organizations in connection with the institutional arrangements they promote.[48] A cursory examination of the long history of colonial

45. For a characteristic formulation see Robert H. Haveman, *The Economics of the Public Sector*, 2d ed. (Santa Barbara: Wiley, 1976), especially pt. 2.

46. Wolf, "A Theory of Nonmarket Failure." For an intriguing argument that focuses on the growing rigidities of aging organizations see Mancur Olson, Jr., *The Rise and Decline of Nations* (New Haven: Yale University Press, 1982).

47. See also the elegant argument presented in Albert Hirschman, *Shifting Involvements: Private Interest and Public Action* (Princeton: Princeton University Press, 1982).

48. On the concept of hegemony consult Robert O. Keohane, *After Hegemony: Discord and Collaboration in the World Political Economy* (Princeton: Princeton University Press, 1984), chap. 3. For an account that explicitly adopts the perspective of the hegemon see

regimes offers some support for this point of view.[49] A major theme of colonial history, after all, is the preference of imperial powers for networks of bilateral relations in which individual colonies interact on a superordinate/subordinate basis with imperial powers but have little opportunity to interact with each other. The essential attraction of such arrangements is that they facilitate the efforts of imperial powers to maintain control by isolating colonial leaders and minimizing opportunities for colonies to band together in opposition to the policies of imperial states. Naturally enough, such networks of bilateral relations do not generate significant roles for central organizations. Yet it is worth noting that even colonial regimes sometimes give rise to central organizations. Perhaps the classic example, though certainly not the only case in point, is the British Commonwealth of Nations.[50] The Commonwealth evolved after World War I as a means of holding together a colonial regime in the face of progressive erosion in the physical and moral dominance of Great Britain as an imperial state. Operating organizationally through the Commonwealth Secretariat, it served to co-opt numerous leaders of individual colonies into the colonial order by giving them stakes in the perpetuation of the system. The struggles of many Indian leaders in the 1940s over participation in the Commonwealth as well as the continuing ties of states like Canada and New Zealand to the Commonwealth attest to the remarkable effectiveness of this arrangement and the organization to which it gave rise.

When we turn to other sorts of imposed regimes, the attractions of central organizations loom even larger. In some cases, organizations appeal to hegemons seeking to distract attention from their dominant positions and to engender widespread feelings of legitimacy regarding the institutional arrangements they create. The IMF originated in this sort of environment. The United States (or perhaps the United States in consort with Great Britain) dominated the process of creating the postwar international monetary regime.[51] Yet the United States chose to deemphasize its hegemonic role in this context, a fact that made the creation of an organization like the IMF appealing. An interesting variation on this theme involves the establishment in 1948 of the Organiza-

---

Arthur A. Stein, "The Hegemon's Dilemma: Great Britain, the United States, and the International Economic Order," *International Organization* 38 (1984), 355–386.

49. For an informative account of the imperial mentality see A. P. Thornton, *The Imperial Idea and Its Enemies* (Garden City: Doubleday, 1968).

50. Hedley Bull, "What is the Commonwealth?" in *International Political Communities* (Garden City: Doubleday, 1966), 457–468.

51. Richard N. Gardner, *Sterling-Dollar Diplomacy in Current Perspective* (New York: Columbia University Press, 1980).

tion for European Economic Cooperation (OEEC) in connection with the European Relief Program (or Marshall Plan). Surely, the United States played a preponderant role in launching the postwar regime of economic cooperation in Western Europe, which, in turn, gave rise to the Common Market. Realizing that it would not do to attempt to supervise this arrangement too closely, however, the United States deliberately promoted the creation of the OEEC as a condition for the provision of economic aid to individual European states.

In other cases, organizations play more pragmatic roles in conjunction with imposed regimes. A case in point is the Council for Mutual Economic Assistance (CMEA) in Eastern Europe. The contemporary regime in Eastern Europe, which originated with Soviet military occupation at the close of World War II, has long since evolved into an arrangement under which the Soviet Union, in effect, obtains political allegiance on the part of Eastern European states in exchange for economic subsidies (for example, secure supplies of energy at lower than world market prices).[52] To be sure, this arrangement is backed by the implicit threat of Soviet military intervention. But this does not diminish the importance of the role played by the organizational structure of the CMEA in making the regime work.[53] Somewhat similar remarks are probably in order regarding the North Atlantic Treaty Organization (NATO), at least during the early years when the dominance of the United States was indisputable. A major function of NATO has always been to promote burden or cost sharing in an environment in which there is a pronounced incentive for NATO members other than the United States to treat NATO as a privileged group in which the United States will continue to supply the public good of collective defense even if they act as free riders.[54]

Since the creation of organizations always involves conscious planning, it is hard to envision a role for organizations of any kind in conjunction with spontaneous regimes. And the evidence suggests that this expectation is a fair characterization of the spontaneous regimes that have arisen in international society. Even so, a little discussion will reveal at least a partial exception to this general proposition.

In his influential analysis of the evolution of cooperation, Axelrod explores the processes through which spontaneous or, in Hayek's phrase,

52. Michael Marrese, "CMEA: Effective but Cumbersome Political Economy," *International Organization* 40 (1986), 287–327.

53. Andrzej Korbonski, "COMECON: The Evolution of COMECON," in *International Political Communities,* 351–403.

54. See also Mancur Olson, Jr., and Richard Zeckhauser, "An Economic Theory of Alliances," *Review of Economics and Statistics* 48 (1966), 266–279.

self-generating regimes are likely to emerge.[55] In fact, he sets out explicitly to investigate the conditions under which cooperation will "emerge in a world of egoists without central authority," and he presents a convincing argument to the effect that the relevant conditions are not farfetched.[56] This line of reasoning undoubtedly helps us to comprehend why it is a mistake to confuse many of the anarchical arrangements prevailing in international society today with situations characterized by disorder or chaos. In the process, it also reinforces the views of those who cherish freedom and abhor the consequences of nonmarket failures and who, therefore, are attracted to institutional arrangements that circumvent the need for organizations altogether or, at least, minimize the role of organizations.

Yet even regimes that arise spontaneously often evolve over time in such a way that participants in the resultant social practices acquire powerful incentives to establish organizations to protect these arrangements from destabilizing pressures. Open-to-entry common-property regimes for the marine fisheries arose spontaneously and worked reasonably well prior to the development of modern technologies (for example, the high-endurance stern trawler) for high-seas fishing. But the advent of these technologies quickly led to radical instabilities and pressures to reform the relevant regimes on a conscious basis.[57] One common response has been to add organizations to regimes of this sort (for example, the International North Pacific Fisheries Commission, the Northwest Atlantic Fisheries Organization, the Inter-American Tropical Tuna Commission) as part of a move to transform them from open-to-entry into restricted common-property regimes.[58] Similar observations are in order regarding certain exchange systems in which competitive markets emerged on a more or less spontaneous basis. Thus, the growing instabilities of many commodities markets have led to the creation of organizations concerned with sugar, coffee, rubber, and tin as part of revised arrangements intended to manage the markets for these commodities. The track record of these organizations has not been outstanding.[59] And, of course, it is reasonable to conclude that the addition of

55. Robert Axelrod, *The Evolution of Cooperation* (New York: Basic Books, 1984), and Friedrich A. Hayek, *The Political Order of a Free People*, vol. 3 of *Law, Legislation, and Liberty* (Chicago: University of Chicago Press, 1979).

56. Ibid., 3.

57. For a rich descriptive account of this process in the case of the North Atlantic fisheries see William Warner, *Distant Water* (Boston: Little, Brown, 1983).

58. Charles B. Heck, "Collective Arrangements for Managing Ocean Fisheries," *International Organization* 29 (1975), 711–743.

59. Vincent A. Mahler, "The Political Economy of North-South Commodity Bargaining: The Case of the International Sugar Agreement," *International Organization* 38 (1984),

central organizations transforms these arrangements from spontaneous regimes into negotiated regimes. But it is interesting to consider dynamic developments of this sort, which stimulate participants to create organizations even in conjunction with institutional arrangements that originated as spontaneous regimes.

FREESTANDING ORGANIZATIONS

Initially, the idea of an organization operating in the absence of a distinct relationship to an institutional arrangement seems anomalous. It conjures up an image of an administrative apparatus with nothing to administer. And there is little evidence of international organizations operating without reference to the broad institutional arrangements existent in international society, such as the states system or the international economic order. Yet organizations that are freestanding in the sense that they are not intimately linked to specific regimes are plentiful in international society, just as they are in domestic societies. A number of the specialized agencies of the United Nations system, such as the Food and Agriculture Organization, the World Health Organization, and the World Meteorological Organization, are fundamentally freestanding organizations in these terms. So also are many organizations not linked to larger organizational structures, such as the International Council for the Exploration of the Sea (whose inaugural meeting took place in 1902), the Intergovernmental Oceanographic Commission, and the International Centre for the Study of the Preservation and Restoration of Cultural Property.

Some freestanding organizations specialize in service activities. They endeavor to provide well-defined services of value to various groups in international society without becoming enmeshed in the larger controversies or conflicts that divide the members. The United Nations Institute for Training and Research conducts research on subjects of interest to its members. The International Institute for Applied Systems Analysis, created in 1972 by twelve national academies of science (including Eastern European as well as Western groups), seeks to bring science to bear to promote international cooperation. The International Council of Scientific Unions (ICSU) coordinates the activities of a collec-

709–731; Robert H. Bates and Da-Hsiang Donald Lien, "On the Operations of the International Coffee Agreement," *International Organization* 39 (1985), 553–559; and Mark W. Zacher, "Trade Gaps, Analytical Gaps: Regime Analysis and International Commodity Trade Regulation," *International Organization* 41 (1987), 173–202.

tion of more specific organizations that focus on individual scientific disciplines. The Scientific Committee for Antarctic Research (SCAR), which has become a major player in coordinating the scientific research programs of various countries active in Antarctica, is actually a committee of ICSU and not an organization with any formal link to the international regime for Antarctica.[60] The International Bureau for Whaling Statistics, which performs a valuable service by assembling and disseminating information on whale harvests, is not formally a part of the international whaling regime. The International Institute for Strategic Studies compiles and disseminates unbiased information on the global strategic balance and the military activities of individual states. As these examples suggest, the collection of freestanding service organizations active in international society today is not only large but also highly varied in functional terms.

Other freestanding organizations are better understood as problem-solving agencies in that they seek to contribute to the amelioration of divisive issues. Several agencies associated with the United Nations system, such as UNESCO and the Food and Agriculture Organization, played significant parts in facilitating the international process of decolonization in the postwar era. Other organizations, like the International Peace Academy, have sought to augment international capabilities in the realm of conflict management and peacekeeping. Still others, such as the United Nations Industrial Development Organization and the United Nations Development Program, have become active in the effort to devise successful strategies for economic development in the less developed countries. Amnesty International, a nongovernmental organization, strives to alleviate human rights violations in a wide range of countries. Another nongovernmental organization, the World Council of Indigenous Peoples, focuses on the problems of aboriginal peoples (known collectively as the Fourth World) locked into states they cannot hope to control. In all these cases, freestanding organizations seek to ameliorate well-defined problems rather than to assume any major role in restructuring the institutional arrangements prevailing in international society.

By contrast, a third category of freestanding organizations encompasses those that are overtly regime oriented in that one of their major purposes is to contribute to the formation of new international institutions. The United Nations Conference on Trade and Development, for

60. On the role of SCAR see James H. Zumberge, "The Antarctic Treaty as a Scientific Mechanism: The Scientific Committee on Antarctic Research and the Antarctic Treaty System," in Polar Research Board, *Antarctic Treaty System: An Assessment* (Washington, D.C.: National Academy Press, 1986), 153–168.

instance, has become a major forum for those desiring to promote the package of alterations in the prevailing international economic order that is known collectively as the new international economic order. More concretely, the United Nations Environment Programme, which was created in the wake of the 1982 Stockholm Conference on the Human Environment but which is not actually linked to an existing regime, has played a remarkably effective role in promoting the development of geographically defined regimes that focus on pollution control under the terms of its Regional Seas Programme[61] as well as in facilitating the negotiation of a more general regime to protect stratospheric ozone. The International Maritime Organization has provided an important forum for the development of an evolving regime that deals with the pollution of the sea from ships. This regime rests on an interlocking set of international conventions that culminated in the provisions of the 1978 International Convention on the Prevention of Pollution from Ships.[62] Similar observations are in order regarding the role of the International Atomic Energy Agency in encouraging the negotiation of an international regime to cope with the problems associated with radioactive fallout. The Antarctic and Southern Ocean Coalition, a nongovernmental organization, has succeeded in playing a part both in protecting the existing regime for Antarctica and in devising supplemental regimes for marine living resources and minerals in the Antarctic region. Though it is true that regimes emerge spontaneously under some conditions, conscious efforts to create institutional arrangements are frequently protracted and difficult. As a result, freestanding organizations often become not only sources of innovative ideas for those negotiating regimes but also pressure groups and even negotiating forums in the lengthy processes involved in the creation of negotiated regimes. In such cases, freestanding organizations constitute an advance guard that endeavors to move international society toward the development of a richer texture of institutional arrangements.[63]

INTERNATIONAL STATE OF NATURE

We come, finally, to the condition I have labeled an international state of nature. This condition encompasses any situation characterized by

---

61. Caldwell, *International Environmental Policy,* 129–142.

62. R. Michael M'Gonigle and Mark W. Zacher, *Pollution, Politics, and International Law: Tankers at Sea* (Berkeley: University of California Press, 1979), and Robin Churchill, "The Role of IMCO in Combating Marine Pollution," in Douglas J. Cusine and John P. Grant, eds., *The Impact of Marine Pollution* (London: Croom Helm, 1980), 73–94.

63. Note also that organizations may play a rearguard role when they survive the

the absence of both institutions and organizations. In a state of nature, individual actors are free to do as they please, without any pressure to comply with institutionalized rules, seek the consent of others, or even consider the consequences of their actions for others.

In one sense, there are no states of nature in contemporary international society.[64] The states system establishes a comprehensive framework of rules or conventions applicable to the realm of terrestrial affairs. The emerging orders for the oceans and space complement the states system in the realm of nonterrestrial affairs. Yet there are certainly many functionally specific areas of international relations not covered by specialized international regimes and not a matter of concern for international organizations. Antarctica, prior to the establishment of the regime articulated in the 1959 treaty, exemplified this condition. There is, today, no comprehensive regime or organization oriented to the management of Arctic issues.[65] Much the same can be said of many large natural systems. The regime for outer space is rudimentary at best, and no regime has arisen so far to structure the activities of the members of international society affecting the global climate system.[66] In this more limited sense, it seems appropriate to speak of conditions approximating an international state of nature.

In fact, such conditions are perfectly tolerable so long as levels of human activity remain low and there is little interdependence among parties active in a given area. To put it another way, institutional arrangements generally arise to lend order to human activities and, especially, to regulate conflicts of interest arising in connection with human endeavors. There was no need for a regime covering Antarctica before states began to assert jurisdictional claims to the continent and to establish research stations in the region. No one felt the absence of a regime for the Arctic Basin so long as the region was of interest only to a handful of explorers and scientists together with a small population of indigenous peoples living in harmony with the natural environment. Similarly, it would have made little sense to worry about a regime for outer space or celestial bodies before technology advanced far enough to permit human beings to engage in activities having significant impacts on these areas. The same is true of the upper atmosphere and the global climate system, now increasingly threatened by such problems as the depletion

---

regimes they were associated with to become, in effect, freestanding organizations. Some of the regional fisheries organizations appear to be heading in this direction.

64. Bull, *The Anarchical Society*.

65. See the collection of essays on polar politics published in *International Journal* 39 (Autumn 1984).

66. Edith Brown Weiss, "International Responses to Weather Modification," *International Organization* 29 (1975), 805–826, and Brown et al., *Regimes for the Ocean*.

of stratospheric ozone and the greenhouse effect. As these examples also suggest, however, the expansion of human capabilities gives rise to an ever-increasing need for institutional arrangements at the international level as well as at the domestic level. With the growth of human activities comes the prospect of severe disruptions of natural systems and serious social conflict. The fundamental purpose of institutions is to control these products of human activities without suppressing the activities themselves.

LESSONS FOR INSTITUTIONAL DESIGN

Perhaps it will help to conclude this chapter by highlighting the principal implications of the preceding analysis of relationships between international institutions and international organizations for efforts to cope with emerging issues in international society.[67] For simplicity, I shall organize this account in the form of a series of propositions directed toward policymakers endeavoring to come to terms with issues of institutional design.[68]

1. *The distinction between institutions and organizations is fundamental.* There are roles for both institutions and organizations in international society. But the common practice of using these terms interchangeably not only sows confusion, it also impedes efforts to think creatively about appropriate combinations of institutions and organizations at the international level.

2. *Some problems associated with the operation of institutions can be solved without resorting to the establishment of organizations.* While organizations can play a variety of roles, it is easy to exaggerate the need for organizations to administer institutional arrangements. Organizations are costly both in strictly monetary terms and in more intangible terms. Accordingly, there should be no bias against anarchical arrangements or presumption in favor of institutions that feature elaborate organizations. Policymakers should strive to create organizations only when there is a demonstrable need for them.

3. *When policymakers do choose to establish organizations, they should tailor these arrangements to the characteristics of the institution or regime under consideration.* Organizations are not appropriate or inappropriate in their own

67. For an account stressing the significance of this perspective see Kratochwil and Ruggie, "International Organization."
68. See also the observations in Jan Tinbergen, "Alternative Forms of International Cooperation: Comparing Their Efficiency," *International Social Science Journal* 30 (1978), 223–237.

right. Rather, they are more or less well suited to the institutional arrangements with which they are associated. It follows that policymakers should design organizations to maximize their effectiveness and minimize their negative by-products in connection with the provisions of specific institutions or regimes.

4. *Policymakers should avoid falling prey to intellectual fashions in thinking about the linkages between institutions and organizations.* Throughout much of the postwar period, observers treated international anarchy as inferior to civil society. Today, we are witnessing a marked increase of emphasis on nonmarket failures leading to minimalist views regarding roles for organizations at the international level, not to mention governments at the domestic level. Yet neither of these intellectual fashions offers a trustworthy guide for policymakers concerned with institutional design. There is no substitute for starting with a systematic assessment of the relative merits of alternative institutional arrangements and proceeding to devise organizations (if any) to meet the requirements of the specific institutional arrangements selected.

CHAPTER THREE

The Power of Institutions: Why International Regimes Matter

Students of international relations often differ sharply from students of domestic society in the assumptions they make about the significance of social institutions. Most observers of domestic society simply take it for granted that institutions such as structures of property rights, markets, electoral systems, or mechanisms for settling disputes are major determinants of collective outcomes in human affairs.[1] They even ascribe a coercive quality to social institutions in that they see subjects as experiencing powerful pressures to conform to the dictates of institutional arrangements, whether they like it or not. To question these assumptions would be to challenge some of the most basic premises of modern, Western thought on the operation of human social systems. Orthodox students of international relations, by contrast, typically take the view that social institutions are of little or no significance as determinants of collective outcomes in international society. Rather, they assume that international institutions, including regimes of various sorts, are mere surface reflections of underlying forces or processes, subject to change with every shift in the real determinants of collective outcomes.[2]

What accounts for this striking difference? More to the point, is there a persuasive justification for this dichotomy? In this chapter, I argue that the difference arises, fundamentally, from certain misunderstandings of the nature of social institutions and a mistaken conception embedded in

---

1. This perspective is exemplified with particular clarity in the public choice literature. See, for example, James M. Buchanan and Gordon Tullock, *The Calculus of Consent* (Ann Arbor: University of Michigan Press, 1962). And for an account that probes the underpinnings of this perspective consult Geoffrey Brennan and James N. Buchanan, *The Reason of Rules: Constitutional Political Economy* (Cambridge: Cambridge University Press, 1985).

2. Susan Strange, "*Cave! hic dragones:* A Critique of Regime Analysis," in Stephen D. Krasner, ed., *International Regimes* (Ithaca: Cornell University Press, 1983), 337–354.

the orthodox account of international relations, of the circumstances facing individual members of international society. I endeavor, in the process, to offer a convincing explanation of why institutions matter in international society just as they do in domestic society.

The Orthodox Account

The orthodox account of international relations contrasts international society and domestic society. Some of the key differences are structural in that they pertain to attributes of the international system rather than to characteristics of the individual members of international society.[3] Above all, international society is organized in an anarchical fashion.[4] No hierarchy of authority exists among its members. There is no central government capable of making binding decisions about the common good and bringing pressure to bear on individual members to comply with these decisions. Individual members cannot rely on centralized enforcement agencies to compel compliance with international rules or conventions, a fact that has led some observers to describe international society as a self-help system and persuaded others to question whether international law should be treated as law at all.[5] Additionally, international society exhibits a comparative lack of the division of labor among individual members that has become pervasive in most domestic societies. States typically endeavor to function as generalists at the international level, maintaining a sense of self-sufficiency rather than becoming specialists in the manner of individuals in domestic society.[6] In Durkheim's terms, therefore, international society can exhibit only a mechanical social order in contrast to an organic social order.[7] Accordingly, the constraints associated with high levels of mutual dependence will be less prevalent at the international level than they are at the domestic level.[8]

3. For an account that stresses these structural considerations see Kenneth N. Waltz, *Theory of International Politics* (Reading: Addison-Wesley, 1979).

4. Hedley Bull, *The Anarchical Society* (New York: Columbia University Press, 1977).

5. For a sophisticated discussion see Anthony D'Amato, "What 'Counts' as Law?" in Nicholas Greenwood Onuf, ed., *Lawmaking in the Global Community* (Durham: Carolina Academic Press, 1982), 83–107.

6. Kenneth N. Waltz, "Reflections on 'Theory of International Politics': A Response to My Critics," in Robert O. Keohane, ed., *Neorealism and Its Critics* (New York: Columbia University Press, 1986), 322–345.

7. Emile Durkheim, *The Division of Labor in Society* (New York: Free Press, 1983/1964).

8. Even so, some observers have taken to stressing the rising level of interdependence at the international level. See, for example, Robert O. Keohane and Joseph S. Nye, *Power and Interdependence: World Politics in Transition* (Boston: Little, Brown, 1977).

Other key differences, according to the orthodox account, pertain to the units, or members, of international society in contrast to the individuals who populate domestic societies. Sovereign states are the primary units of international society. Unlike individuals at the domestic level, sovereign states are autonomous in that they are not subject to restrictions or constraints articulated and imposed in the name of the common good.[9] Equally important, states are driven by the pursuit of a relatively narrow form of self-interest defined largely in terms of power. Unlike individuals, states seldom behave in an altruistic fashion or respond to the dictates of moral or ethical standards.[10] In fact, they are not even particularly sensitive to the benefits of enlightened self-interest. For the most part, sovereign states simply pursue their immediate interests defined, more often than not, in terms of the accumulation or use of power.

The significance of these differences, according to the orthodox account, is straightforward. Power politics constitute the principal determinant of collective outcomes in international society in that outcomes flow directly from interactions among autonomous states, each of which is attempting to maximize its own power.[11] It follows that those who wish to explain collective outcomes at the international level must endeavor to comprehend the nature of power, the bases of power in international society, the strategies deployed by those seeking power, and the dynamics of the interactions among states endeavoring to maximize their power.[12] Of course, this endeavor encompasses the study of alliances or coalitions as well as the actions of states operating on an individualistic basis. Likewise, explanations of major changes in international relations must be sought in analyses of shifts in the configuration of power in international society.

On this account, institutions are mere surface reflections of underlying processes that involve the dynamics of power.[13] They may serve as indicators of bargains struck among sovereign states or the ascendance of a particular state to a position of dominance.[14] But they cannot

9. Hans J. Morgenthau, *Politics among Nations* (New York: Knopf, 1948).
10. But see the interesting observations in Charles P. Kindleberger, "Hierarchy versus Inertial Cooperation," *International Organization* 40 (1986), 841–847.
11. Morgenthau, *Politics among Nations*.
12. For a review of the conceptual and analytic problems associated with the study of power consult David A. Baldwin, "Power Analysis and World Politics: New Trends versus Old Tendencies," *World Politics* 31 (1979), 161–194.
13. As Strange puts it, "All those international arrangements dignified by the label regime are only too easily upset when either the balance of bargaining power or the perception of national interest (or both together) change among those states who negotiate them" ("*Cave! hic dragones*," 345).
14. This view seems to underpin a substantial portion of the analysis in Robert O.

operate as determinants of collective outcomes at the international level in their own right. Order, if any emerges, in international society arises either from the operation of an effective balance of power or from the emergence of a hegemon or dominant power in the system.[15] A balance of power may lend an element of stability to international relations by providing an effective set of checks and balances.[16] Under other circumstances, a hegemon may conclude that its own interests will be well served by imposing certain patterns of behavior on the other members of international society.[17] In the end, however, we must remember that order stems from the configuration of power in the system rather than from any constraints associated with the presence of social institutions.

Where does this leave such international institutions as the postwar international monetary regime, the liberal regime for international trade, or the emerging arrangements for the oceans and outer space?[18] Whereas students of domestic society regularly speak of institutional constraints on the actions of individuals, orthodox students of international relations treat institutions as epiphenomena. Given the paradigm described in the preceding paragraphs, the reasoning underlying this point of view is easy to describe. States can simply discard existing international institutions when they become inconvenient or troublesome and replace them with alternative arrangements that are more to their liking. Since institutions are only surface reflections of underlying power relationships, there is nothing to stop states from proceeding in this fashion. What is more, individual states can ignore the dictates of institutions with impunity, even when it seems convenient to retain existing arrangements on a pro forma basis. In an anarchical system like international society compliance is voluntary, and there is nothing to stop individual members from violating rights or rules whenever the pursuit of power calls for such a course. It follows that institutions cannot be important determinants of collective outcomes in international society. This does not mean that international institutions are of no interest to orthodox students of international relations. Since trans-

Keohane, *After Hegemony: Cooperation and Discord in the World Political Economy* (Princeton: Princeton University Press, 1984).

15. Inis L. Claude, Jr., *Power and International Relations* (New York: Random House, 1962).

16. Ernst B. Haas, "The Balance of Power: Prescription, Concept, or Propaganda?" *World Politics* (1953), 442–477.

17. Keohane, *After Hegemony*, especially chap. 3.

18. For discussions of these regimes consult Krasner, *International Regimes;* Seyom Brown, Nina W. Cornell, Larry L. Fabian, and Edith Brown Weiss, *Regimes for the Ocean, Outer Space, and Weather* (Washington, D.C.: Brookings, 1977); and Kenneth A. Oye, ed., *Cooperation under Anarchy* (Princeton: Princeton University Press, 1986).

formations of institutional arrangements generally reflect shifts in the configuration of power in the system, it is worth keeping track of institutional changes as a means of gauging the evolution of the structure of power in international society.[19] But those who seek to explain collective outcomes at the international level will not get far by directing their attention to international institutions.

The central argument of this chapter begins here. In the following sections, I endeavor to demonstrate that the orthodox account's relegation of international institutions to the status of epiphenomena is mistaken on several counts. Above all, it rests on an erroneous conception of the circumstances facing individual members of international society day to day. From the perspective of the individual state, international institutions are, for the most part, facts of life. States sometimes have the option of participating or refusing to participate in institutional arrangements. They may, for example, choose to opt out of institutions like the international monetary regime or the new regime governing the deep seabed. In general, however, it is exceedingly difficult to get rid of international institutions and to replace them with more attractive alternatives. If anything, it is harder to replace international institutions than to replace institutional arrangements in domestic society. Additionally, there are usually substantial costs involved in ignoring the dictates of international institutions, even in the absence of explicit sanctions at the disposal of authorized enforcement agencies. Like individuals in domestic society, states generally comply with the rights and rules of international institutions voluntarily if only in the sense that they find that the benefits of compliance outweigh the benefits of violation, whether or not formal sanctions figure in their calculations.

It follows, as I shall argue, that the orthodox contrasts between domestic society and international society are greatly exaggerated, at least when it comes to evaluating the significance of social institutions. Just as institutional arrangements appear to be major determinants of collective behavior at the domestic level, there is every reason to expect them to play a similar role at the international level.

THE DURABILITY OF INTERNATIONAL INSTITUTIONS

If the members of international society could simply discard institutional arrangements that seem inconvenient or troublesome, it would be difficult to avoid the conclusion that regimes, or international institu-

19. See also Keohane and Nye, *Power and Interdependence*, especially pt. 2.

tions more broadly, are epiphenomena of little significance as determinants of collective outcomes. To demonstrate the significance of international institutions, therefore, it is necessary to show that the members of international society cannot get rid of them easily. This, in turn, requires an examination of institutional dynamics at the international level.

As in all social systems, institutional arrangements in international society change continually. Sometimes they even undergo rapid change. Consider, for example, the transformations in monetary regimes that have taken us from the gold standard of the early twentieth century through the era of fixed exchange rates under the Bretton Woods system to the floating exchange rates that govern relationships among national currencies today.[20] Witness also the breakup of the nineteenth-century version of colonialism that eventuated in the virtual disappearance of this institution in the years following World War II.[21] Or recall the rapid transition during the postwar era from relatively unrestricted common-property arrangements for marine areas to regimes featuring coastal-state jurisdiction in the form of exclusive economic zones for large areas of coastal waters and regimes featuring the principle of the common heritage of mankind for areas remaining in the international commons.[22]

In exceptional cases, a dominant state (that is, a hegemon) or a powerful coalition (that is, an international club) can assume leadership by giving conscious direction to transitions in institutional arrangements.[23] Unsettled periods following the disruptions associated with major wars or world depressions undoubtedly offer striking illustrations of this phenomenon. We are all familiar with the remarkable role of the United States in devising new international trade and monetary regimes during the 1940s in the aftermath of the Great Depression and World War II.[24] But leadership in international-regime formation is not limited to the aftermath of major disruptions. The twelve original signatories to the

20. Benjamin J. Cohen, "Balance-of-payments Financing: Evolution of a Regime," in Krasner, *International Regimes*, 315–336. For additional details consult Benjamin J. Cohen, *In Whose Interest? International Banking and American Foreign Policy* (New Haven: Yale University Press, 1986).

21. Rupert Emerson, *From Empire to Nation* (Cambridge: Harvard University Press, 1960).

22. Ross D. Eckert, *The Enclosure of Ocean Resources* (Stanford: Hoover Institution Press, 1979), and Bernard H. Oxman, David D. Caron, and Charles L. O. Buderi, eds., *Law of the Sea: U.S. Policy Dilemma* (San Francisco: ICS Press, 1983).

23. On the distinction between leadership and hegemony, as the terms are commonly used, see the insightful observations in Kindleberger, "Hierarchy versus Inertial Cooperation."

24. For a brief account see Keohane, *After Hegemony*.

Antarctic Treaty of 1959 (that is, the Antarctic club) certainly played a leadership role in the creation of the current international regime for Antarctica.[25] In much the same way, the United States and a few other like-minded members of international society undoubtedly took the lead in devising the regime governing international trade in endangered species, which is now formalized in the 1973 Convention on International Trade in Endangered Species of Fauna and Flora.[26]

Yet it is misleading to direct attention primarily to these exceptional situations. In the typical case, international institutions are extremely difficult to get rid of, or even to adjust in a conscious fashion to changing circumstances. Interestingly, existing regimes or institutional arrangements often prove highly resistant even to assaults spearheaded by one or more of the great powers. Common-property regimes for renewable resources, for example, are remarkably difficult to change at the international level. Even the enclosure movement, which has brought many resources within the exclusive management authority of coastal states, has done little to eliminate the common-property character of regimes for marine fisheries and other renewable resources of marine regions.[27] The obvious importance, under contemporary conditions, of moving toward regimes based on an ecosystems approach or at least a multi-species approach at the international level has so far made little headway against the persistent tendency to develop separate regimes for individual species, such as polar bears, fur seals, or halibut.[28] The efforts of those endeavoring to make significant changes in the existing international economic order, such as the various reforms often lumped together under the general rubric of a new international economic order, have met with remarkable resistance despite the fact that the reform movement probably includes a majority of the members of international society.[29] To be more specific, the Soviet Union failed completely in its attempt to promote major changes in the international regime governing the Svalbard Archipelago during the 1940s.[30] Similarly, the efforts of a large coalition of states to reform the prevailing international re-

25. Philip W. Quigg, *A Pole Apart* (New York: McGraw Hill, 1983).

26. Lynton Keith Caldwell, *International Environmental Policy* (Durham: Duke University Press, 1984), especially 189–192, and Simon Lyster, *International Wildlife Law* (Cambridge: Grotius Publications, 1985), chap. 12.

27. Oran R. Young, "The Political Economy of Fish: The Fishery Conservation and Management Act of 1976," *Ocean Development and International Law* 10 (1982), 199–273.

28. For an account of efforts to promote an ecosystems approach see Caldwell, *International Environmental Policy*, chaps. 7 and 8.

29. Marvin S. Soroos, *Beyond Sovereignty: The Challenge of Global Policy* (Columbia: University of South Carolina Press, 1986), chap. 6.

30. Willy Ostreng, *Politics at High Latitudes* (London: C. Hurst, 1977), especially chap. 6.

gime for Antarctica have met with so much opposition from the members of the Antarctic club that some of the reformers (for example, India) have decided to join the club rather than to continue pushing for changes from the outside.[31]

Perhaps even more striking is the resistance of many international institutions to change even when the prevailing arrangements are not only inefficient but also exceedingly difficult to justify on ethical grounds. The international food regime, which is based on a system of distribution through world markets (in other words, of withholding food from those who lack the money to pay for it), has played a major role in bringing about the current situation in which millions of people are malnourished, in some cases to the point of starvation, despite the fact that global food production is entirely adequate to feed all the world's people. And no serious prospects for reforming this regime are in sight at the present time.[32] While some efforts are now being made to protect living resources in the face of a rising tide of extinctions of species, those who have tried can attest to the difficulties encountered in efforts to restructure institutional arrangements in this area, particularly when it comes to arrangements for noncommercial species such as oceanic birds, porpoises, or sea lions.[33] Much the same can be said of the relatively unrestricted common-property regimes currently in place for the international atmospheric commons. While evidence mounts concerning threats to the world's climate system associated with the buildup of carbon dioxide in the atmosphere or the depletion of ozone in the stratosphere, efforts to devise new institutional arrangements at the international level to cope with these problems are still in their infancy.[34]

There are several clear-cut reasons why international institutions, like domestic institutions, are generally difficult to get rid of or to replace with more appropriate alternatives. Above all, efforts to reform institutional arrangements on an intentional or planned basis almost always entail processes of collective choice. The image of a dominant state or a hegemon playing the role of lawgiver is severely distorted. Even under the extraordinary conditions that prevailed in the aftermath of World

31. Deborah Shapley, *The Seventh Continent: Antarctica in a Resource Age* (Washington, D.C.: Resources for the Future, 1986).

32. See also Raymond F. Hopkins and Donald J. Puchala, "Perspectives on the International Relations of Food," *International Organization* 32 (1978), 581–616.

33. On the problem of disappearing species see Norman Myers, *The Sinking Ark: A New Look at the Problem of Disappearing Species* (New York: Pergamon, 1979).

34. The ozone agreement concluded in Montreal in September 1987 is certainly a significant step. Already, however, a consensus is emerging on the proposition that this agreement does not go far enough toward controlling the production and emission of chlorofluorocarbons.

War II, the United States had to negotiate with several other major players to achieve consensus on the monetary regime set forth in the Bretton Woods agreement. Likewise, the United States found itself forced to make substantial concessions regarding the international-trade regime after the collapse of arrangements laid out in the Havana Charter and the failure to reach consensus on the proposed International Trade Organization.[35] Under more typical conditions, such as those prevailing today, it is clear that regime formation almost invariably gives rise to extremely complex processes of collective choice, which will often fail to produce clear-cut agreement on new or revised institutional arrangements. Witness recent efforts to devise suitable institutional arrangements to govern deep-seabed mining, international broadcasting via satellite, threats to the international climate system, or exploratory activities aimed at assessing the mineral potential of Antarctica.[36]

The fact that the development of international institutions typically involves complex collective-choice processes has several important implications for any effort to understand the significance of institutions as determinants of collective outcomes at the international level. In many cases, existing arrangements have vigorous defenders who stand to lose as a result of institutional change, even if the existing arrangements are unattractive from the point of view of certain conceptions of the common good. It is no cause for surprise, for example, to find the United States staunchly supporting the existing economic order in many specific areas, the members of the Antarctic club defending the current regime for Antarctica, or countries where major polluting industries are based resisting changes in international environmental regimes designed to control the long-range transport problems associated with acid deposition or Arctic haze.[37] What is more, even in cases where all parties acknowledge the inadequacy of existing institutional arrangements (for example, international regimes designed to maintain biological diversity by protecting endangered species), complications arising from collective-choice processes can pose severe impediments to institutional reform. In virtually every case, several distinct options are available to those concerned with the replacement of existing institutional arrangements. And it is rare that the preferences of the relevant members of international society fail to differ substantially with regard to these options. The resultant processes of bargaining over the relative attrac-

---

35. See Keohane, *After Hegemony*, pt. 3, for a brief history.

36. On the politics of regime formation see Oran R. Young, "'Arctic Waters': The Politics of Regime Formation," *Ocean Development and International Law* 18 (1987), 101–114.

37. See also the case studies in Soroos, *Beyond Sovereignty*, pt. 2.

tions of alternative points in a contract zone or zone of agreement are often protracted and can easily obstruct efforts to revise institutional arrangements at the international level for considerable periods of time. The extraordinary bargaining associated with the efforts to devise new institutional arrangements for marine areas during the 1970s and 1980s is, of course, still fresh in the minds of most observers.[38] But this case may well come to seem simple by contrast with emerging efforts to devise institutional arrangements to cope with large-scale ecosystems, climate change, or the long-range transport of pollutants.

If anything, the difficulties involved in getting rid of existing institutional arrangements and introducing new arrangements are more severe in international society than they are in domestic society. This is so because domestic societies generally have highly developed mechanisms, in the form of legislatures, whose principal function is to devise and reform institutional arrangements.[39] Of course, legislatures regularly encounter serious problems in their efforts to deal with institutional issues. Witness the recent struggles of the United States Congress to adjust existing regimes that deal with taxes, welfare, the support of agriculture, or the control of acid deposition. Conversely, legislative conferences sometimes prove quite effective in developing new institutional arrangements at the international level.[40] Quite apart from certain highly visible cases like the Bretton Woods Conference or the Law of the Sea Conference, there are many less publicized sessions, such as the World Administrative Radio Conferences (WARC), the institution-building sessions for marine pollution held under the auspices of the International Maritime Organization, and the regime-construction conferences sponsored by the United Nations Environment Programme.[41] Nonetheless, it is hard to escape the conclusion that legislatures in domestic societies typically feature more highly developed procedures for handling institutional changes and can depend on a greater public willingness to accept their conclusions than is true of the legislative conferences that international society relies on to deal with the reform of existing institutional arrangements. It follows that international institutions will often be more difficult to get rid of than domestic institutions, a

38. James K. Sebenius, "Negotiation Arithmetic: Adding and Subtracting Issues and Partics," *International Organization* 37 (1983), 281–316.
39. For an insightful development of this proposition see Friedrich A. Hayek, *The Political Order of a Free People*, vol. 3 of *Law, Legislation, and Liberty* (Chicago: University of Chicago Press, 1979).
40. Soroos, *Beyond Sovereignty*, especially chap. 3.
41. Caldwell, *International Environmental Policy*, and R. Michael M'Gonigle and Mark W. Zacher, *Pollution, Politics, and International Law* (Berkeley: University of California Press, 1979).

conclusion that surely adds weight to the proposition that institutional arrangements are important determinants of collective outcomes at the international level.

We are all aware, as well, that laboriously negotiated revisions in institutional arrangements often prove difficult to implement, fail to achieve the results intended, produce unintended by-products that swamp the effects of the intended results, or are overtaken by changing circumstances before they can be properly instituted. Surely, no one would deny this observation with regard to domestic society. And it is certainly just as applicable at the international level. It is hardly a cause for surprise, for example, that the regime for the deep seabed laid out in Part XI of the Law of the Sea Convention of 1982 remains, for the most part, a paper arrangement. In fact, some observers regard this outcome, at least in part, as a consequence of the character of the proposed regime itself.[42] Similarly, it remains to be seen whether the arrangements envisioned in the Convention on the Conservation of Antarctic Marine Living Resources of 1980 will become realities during the foreseeable future. Accordingly, those desiring to introduce new institutional arrangements must do more than simply obtain the acquiescence of others on paper. And this often proves difficult at the international level, just as it does at the domestic level. Under the circumstances, existing institutional arrangements may prove resistant to real change because entrenched patterns of behavior are hard to dislodge even in the face of a pro forma agreement to discard prevailing arrangements in favor of some superficially preferred alternative.

Then, of course, there are the costs (both monetary and nonmonetary) of discarding existing institutional arrangements and replacing them with preferred alternatives. Those who speak confidently about the ability of the members of international society to abandon institutions that become troublesome or inconvenient regularly overlook the real costs of such a course of action. Even dominant states discover again and again that there are substantial costs of leadership in connection with regime formation. At best, they are apt to find themselves making significant concessions or taking on major burdens in order to get others to acquiesce in their plans for new institutional agreements.[43] And attempts to assert leadership in this realm may prove considerably more costly if they fail to achieve success, thereby revealing a significant

42. Ross D. Eckert, *The Enclosure of Ocean Resources* (Stanford: Hoover Institution Press, 1979).

43. See the discussion of burden sharing in Mancur Olson, Jr., and Richard Zeckhauser, "An Economic Theory of Alliances," *Review of Economics and Statistics* 48 (1966), 266–279.

erosion in the hegemonic stature of the putative leader. In a sense, this fate has befallen the United States in the international monetary regime, despite the fact that it did succeed in precipitating major changes in the regime through unilateral actions in 1971.[44] Even when new institutional arrangements prove generally successful, moreover, the transaction costs associated with institutional reform are virtually always considerable. Witness the time and energy that numerous members of international society have expended on devising new arrangements to govern marine pollution, endangered species, international broadcasting, and human rights. For these reasons, no member of international society is likely to embark lightly on a campaign to replace institutional arrangements. This is not to deny that international institutions do change continually or that the members of international society often endeavor to nudge the course of change in the direction of preferred alternatives. But the image of individual members of international society discarding inconvenient institutions and replacing them with new arrangements on an almost casual basis is surely incorrect.

What are the implications of the argument set forth in the preceding paragraphs for the significance of institutional arrangements as determinants of collective outcomes at the international level? For the most part, the replacement of prevailing arrangements emerges as a protracted, uncertain, and costly process that involves complex interactions with other members of international society. Just as powerful groups do sometimes succeed in exercising considerable influence over the shape of social institutions at the domestic level, dominant states or coalitions occasionally play critical roles in the development of international institutions. But it is easy to become mesmerized by the image of hegemonic power.[45] Circumstances in which hegemons succeed in redesigning international institutions surely constitute the exception rather than the rule, and even dominant states usually find it necessary to engage in elaborate collective-choice processes and to assume substantial burdens of leadership.[46] Much more typical is the case of the individual member that is faced with a choice between participating in prevailing institutional arrangements or becoming an outcast or pariah state. In this connection, what seems remarkable is the extent to which even relatively powerful states espousing radical changes in many international institutions, such as the Soviet Union in the 1920s and 1930s or China in the

44. Cohen, "Balance-of-Payments Financing," and Keohane, *AfterHegemony*.
45. See also the argument developed in Duncan Snidal, "The Limits of Hegemonic Stability Theory," *International Organization* 39 (1985), 579–614.
46. On the distinction between leadership and outright dominance see Kindleberger, "Hierarchy versus Inertial Cooperation."

1950s and 1960s, have gradually come around to the conclusion that it is expedient to join the club and conform to the principal rights and rules of an array of prevailing international institutions.[47]

Under the circumstances, it seems entirely reasonable to treat institutional arrangements as important constraints on the behavior of individual actors in international society, just as we do in thinking about domestic society. To be sure, institutions are not actors in their own right (though the organizations created to administer institutions may sometimes acquire the attributes of actors). Some analysts have jumped quickly from this observation to the conclusion that institutions cannot be significant determinants of collective outcomes in international society.[48] But this is surely incorrect. It is not necessary to be an actor to operate as a determinant of collective behavior at the international or any other level. To think otherwise would require a wholesale revision of our understanding of the role of markets, electoral systems, structures of property rights, and so forth in domestic societies, not to mention our understanding of the impact of systemic or structural factors at the international level.[49]

COMPLIANCE IN INTERNATIONAL SOCIETY

The preceding discussion licenses the conclusion that the members of international society cannot easily discard institutional arrangements when they become inconvenient or troublesome. There remains, however, the notion that individual members can simply ignore the dictates of international institutions in specific situations, even though they may not find it expedient to get rid of the arrangements altogether. To the extent that individual members are able to adopt a posture of this sort toward international institutions, we would once again be forced to conclude that these arrangements are epiphenomena, lacking real significance as determinants of collective outcomes. It is necessary at this juncture, therefore, to turn to an investigation of factors governing compliance with institutional dictates.

Several preliminary observations are in order. Above all, we must avoid imposing a standard of perfection in assessing compliance. No one expects perfect compliance with the rights and rules of institutional

47. On the case of the Soviet Union in the 1920s and 1930s see F. P. Walters, *A History of the League of Nations* (London: Oxford University Press, 1960), especially chap. 30.
48. James N. Rosenau, "Hegemons, Regimes, and Habit-driven Actors," *International Organization* 40 (1986), 849–894.
49. Waltz, "Reflections."

arrangements in domestic society.[50] Even public agencies authorized to operate enforcement mechanisms regularly discover that the marginal costs of obtaining compliance exceed the marginal benefits well before a condition of perfect compliance is reached.[51] Of course, wholesale noncompliance can and sometimes does undermine social institutions, initiating processes leading to institutional change. But we almost always regard a certain level of noncompliance as normal, and we do not expect institutional arrangements to collapse or unravel as a result.[52] Surely, the same situation obtains at the international level. It follows that we should ask whether most members of international society comply with institutional dictates most of the time, rather than seize upon specific instances of noncompliance as evidence for the proposition that institutions are not effective constraints on the behavior of individual actors at the international level.[53]

There is also an important distinction between compliance and enforcement. Enforcement, which involves the actual or threatened use of sanctions by authorized agents of the community to compel individuals to comply with rights and rules, is only one of several bases of compliance from the point of view of the individual members of a social system.[54] In the typical case, in fact, enforcement is a marginal factor in compliance calculations. Most members of social systems comply with the dictates of prevailing institutions most of the time for reasons having little or nothing to do with their expectations regarding the imposition of sanctions.[55] If this were not so, the enforcement mechanisms operated by public authorities would soon find themselves overwhelmed by the problem of compelling reluctant individuals to comply with institutional dictates. Those who emphasize the underdeveloped nature of enforcement mechanisms in international society often lose sight of this fundamental observation. Instead of initiating a broader investigation into the bases of compliance, they exaggerate the contrast between domestic

50. But note that some of those who think in contractarian terms assume perfect compliance for purely analytic reasons. For a case in point see John Rawls, *A Theory of Justice* (Cambridge: Harvard University Press, 1971).

51. Oran R. Young, *Compliance and Public Authority: A Theory with International Applications* (Baltimore: Johns Hopkins University Press, 1979), chap. 7.

52. It is worth noting, however, that the rights and rules of social institutions are not of equal importance to the survival of specific institutional arrangements. That is, levels of violation that we tolerate with respect to some rights and rules might well seem intolerable with respect to others.

53. See also the discussion in Roger Fisher, *Improving Compliance with International Law* (Charlottesville: University Press of Virginia, 1981).

54. Young, *Compliance and Public Authority*.

55. For an insightful discussion of the emergence of norms of cooperation see Robert Axelrod, *The Evolution of Cooperation* (New York: Basic Books, 1984).

society and international society by placing too much weight on enforcement as the key to individual compliance with socially defined rights and rules.[56]

Why do individual actors ordinarily comply with the dictates of institutional arrangements even when they do not expect to face public sanctions for violations? Broadly speaking, the bases of compliance fall into two major categories: decision variables, or considerations that enter into the decision calculus of subjects as unitary actors, and collective-choice considerations, or factors in the internal decision-making processes of the members of international society.

Decision Variables

Many students of international relations have fallen into the habit of treating the benefits produced by social institutions as well as the institutions themselves as public goods.[57] And it is a short step from this perspective to the expectation that the individual members of society will experience powerful incentives to become free riders, enjoying the benefits produced by institutional arrangements while refusing to comply with the dictates of the institutions themselves.[58] But this line of reasoning constitutes a severe distortion of reality at the international level as well as at the domestic level. It turns out, in fact, that many (perhaps most) institutions produce an array of divisible benefits (as well as public goods) and that those who fail to play the game by refusing to comply with institutional dictates must expect to find themselves excluded from some or all of these benefits. Consider some concrete examples of the rewards of participation at the international level. Those who fail to comply with the dictates of the international monetary regime must expect the loss of some or all of their drawing rights as administered by the International Monetary Fund. Members of international shipping regimes who violate the rules pertaining to ship design or operating procedures must anticipate that their flag vessels will be excluded from areas under the jurisdiction of other members. Those who refuse to abide by the rules adopted at WARC sessions cannot expect to enjoy security of tenure with respect to the use of orbital slots for satellites or frequency bands in the electromagnetic spectrum. Un-

56. Political ideology certainly plays a role in structuring thought in this realm. Specifically, there is a definite link between conservatism and the tendency to stress enforcement as a critical factor in compliance decisions.

57. See also the insights laid out in Michael Taylor, *Anarchy and Cooperation* (New York: Wiley, 1976).

58. Olson and Zeckhauser, "An Economic Theory of Alliances."

der the terms of various international insurance schemes, individual members who fail to comply with applicable rules regarding safety and the like may be declared ineligible for the receipt of compensation. Violators of the rules of the international-trade regime may find themselves losing most-favored-nation status in their dealings with other members of the system.[59] Those who refuse to comply with rules pertaining to technology transfer may find that they cannot get others to accept their claims to exclusive use rights in connection with specific mine sites under the terms of the regime for deep-seabed mining.[60] None of this means that individual members will always comply or never conclude that the gains from violation exceed the benefits of compliance. Like their domestic counterparts, however, international institutions often take on a distinctly coercive quality in that the pressures on individual members to comply with their dictates are substantial.

It is fashionable in some circles to describe international society as a self-help system because it lacks public agencies authorized to enforce rights and rules on behalf of the community.[61] But this circumstance does not mean that those who violate the dictates of international institutions can expect to get away with their violations without provoking reactions from other members of the group. On the contrary, various forms of retaliation occur regularly at the international level.[62] Those who impose impermissible restrictions on foreign diplomats residing in their territory must expect others to retaliate in a similar vein against their own diplomats. Violations of the terms of quotas or moratoriums in the international whaling regime are often met with restrictions on the fishing privileges of the offender in the fishery-conservation zones of other states.[63] Individual members who impose tariffs in violation of their obligations under the international-trade regime must anticipate tit-for-tat reactions on the part of other members. Violations of arms-control regimes regularly provoke tit-for-tat retaliations from other participants. Assuredly, retaliation can become a nasty business in decentralized, self-help social systems. In the absence of an effective public authority, it is easy to touch off action/reaction processes in which each participant justifies its own behavior as a necessary response to the

59. Kenneth Dam, *The GATT: Law and International Economic Organization* (New York: Midway Reprint, 1977).

60. See also Lance N. Antrim and James K. Sebenius, "Incentives for Ocean Mining under the Convention," in Oxman, Caron, and Buderi, *Law of the Sea*, 79–99.

61. Waltz, "Reflections."

62. For a sophisticated account, which develops a helpful vocabulary for identifying different types of retaliation, see Axelrod, *Evolution of Cooperation*.

63. The United States, for example, has cut the fishing quotas of Japan, Norway, and the Soviet Union within the American Fishery Conservation Zone for this reason.

provocations of others. Some may see this as a serious weakness of international institutions. Nonetheless, this feature of international society is apt to make individual members think twice before they violate institutional dictates. It takes only a modicum of enlightened self-interest to realize that the cumulative costs of violations may exceed any immediate benefits many times over.

As the preceding paragraph implies, violators of the dictates of institutional arrangements run the risk of touching off uncontrollable processes leading to the disintegration or collapse of complex institutions whose overall worth greatly exceeds any benefits derivable from specific violations. In a sense, this concern is particularly severe in decentralized systems like international society. Those who violate rights and rules in domestic society can take comfort from the fact that authorized agencies will ordinarily intervene to control any ripple effects associated with their violations or to repair the damage by devising new institutional arrangements. In a curious way, they may even approve of the fact that there is some probability that they will be sanctioned for their own violations. At the international level, by contrast, actors must confront the sobering prospect that their violations can lead to the disintegration of important institutional arrangements. We are all familiar with spirals within regimes in connection with tariff wars, competitive currency devaluations, arms races, and escalations of conventional military confrontations. Those who have studied international history or are old enough to remember the collapse of international institutions during the 1930s can hardly fail to appreciate the dangers associated with spirals of this sort.[64] Moreover, we must recognize the impact of demonstration effects in the sense of disruptions to other institutional arrangements arising from the disintegration of specific regimes. It is hard to avoid the conclusion, for example, that the collapse of the old open-to-entry regime for continental-shelf areas during the 1940s played a key role in initiating the subsequent enclosure movement that has transformed a variety of marine regimes.[65] Similarly, a collapse of the international-broadcasting regime might well trigger drastic changes in a number of other regimes that apply to atmospheric resources. Of course, those who wish to revolutionize prevailing institutional arrangements, whether at the international level or at the domestic level, will count on exploiting such demonstration effects to magnify the impact of their efforts to bring

64. Charles P. Kindleberger, *The World in Depression, 1929–1939* (Berkeley: University of California Press, 1973).
65. Eckert, *Enclosure of Ocean Resources,* and Anne L. Hollick, *U.S. Foreign Policy and the Law of the Sea* (Princeton: Princeton University Press, 1981).

about change. But most members of social systems, including international society, will treat the prospect of such demonstration effects as a persuasive reason for exercising considerable caution. In this sense, the interconnectedness of institutions constitutes a conservative force, especially in systems like international society that lack centralized public authorities.[66]

A reputation for trustworthiness is one of the most valuable assets that any member of international society can acquire. Whether we are concerned with credibility in interactions with various partners at a given time or with extended reciprocity in dealings with the same set of partners over time, individual actors will place a high value on maintaining their reputation for trustworthiness.[67] This is particularly true in a social system like international society in which individual members cannot count on authorized agents to compel others to live up to the terms of their agreements or to comply with the prescriptions of regimes they have accepted in principle.[68] Under the circumstances, the costs of becoming stigmatized by others as a rule breaker may be quite severe, as many Third World states have discovered in connection with the dictates of the international monetary and trade regimes. Even a great power, like the United States, is not immune to such considerations. It seems undeniable, for example, that the United States is already paying a price for actions such as its refusal to accept the decision of the International Court of Justice in the Nicaragua harbor-mining case or its attempts to put off the impact of the Court's decision in the Gulf of Maine case on New England fishers. Nor can members of international society hope to safeguard their reputation for trustworthiness by moving unilaterally to change institutional rules as a means of covering up their unwillingness to comply with existing rules. As the behavior of the United States toward the international monetary regime during the 1970s demonstrates, members of international society do adopt such an approach from time to time. But it would be hard to deny that the United States has paid a high price in recent years for this and other unilateral actions, such as its last-minute withdrawal from the effort to reach agreement on new institutional arrangements in the law-of-the-sea negotiations.[69] For

66. See also the analysis of order in international society developed in Friedrich V. Kratochwil, *International Order and Foreign Policy* (Boulder: Westview Press, 1978).

67. Robert O. Keohane, "Reciprocity in International Relations," *International Organization* 40 (1986), 1–27.

68. For a seminal account consult Thomas C. Schelling, *The Strategy of Conflict* (Cambridge: Harvard University Press, 1960).

69. Leigh S. Ratiner, "The Costs of American Rigidity," in Oxman, Caron, and Buderi, *Law of the Sea*, 27–42.

their part, lesser members of international society are unlikely even to perceive such unilateral efforts to alter rules as a live option. Overall, then, those who become known for their violations of the dictates of international institutions must expect to pay a substantial price for their actions in terms of their reputation for trustworthiness, one of the most important assets that a member of international society can acquire.

What is more, the phenomena of social pressure and conformity are not limited to the behavior of individuals in domestic society.[70] Those who direct the external behavior of the members of international society commonly seek acceptance among their peers in other countries. No doubt, we can identify exceptions to this generalization. It seems probable, for example, that Hitler cared little about the reactions he evoked from other statesmen, so long as he was able to advance his material agenda for the Third Reich. Unlike some of the other early Soviet leaders, Trotsky appears to have exhibited little interest in gaining acceptance among other international leaders. Perhaps the same can be said of Mao Tse Tung, especially toward the end of his career. But these cases are surely exceptional. Few leaders feel comfortable in presiding over policies that produce prolonged adverse publicity for their governments or states, must less that eventuate in treatment of their countries as outcast or pariah states. In fact, few leaders can afford the consequences of such policies. This is particularly true of statesmen in "new" states who are concerned to establish their credentials as members of international society in good standing or in states, like Germany and Japan in the aftermath of World War II, whose stature in international society is severely tarnished. In all such cases, leaders experience strong incentives to demonstrate the respectability of their countries by compiling outstanding records of compliance with the rights and rules of international institutions. It is therefore no accident that many Latin Americans and Indians have become well known for their interest in the codification and interpretation of international law. And it is revealing to reflect on the efforts of Soviet leaders to demonstrate their compliance with the dictates of international institutions in the forum of the League of Nations during the 1930s as well as the more recent efforts of Chinese leaders to participate as good citizens in the United Nations. In short, even members of international society that are relatively self-sufficient in socioeconomic terms are not immune to the influence of social pressure. Few leaders find it comfortable or, for that matter, politically safe, to preside over outcast or pariah states, a condition that

70. For a general account of the literature on conformity see C. A. Kiesler and Sara B. Kiesler, *Conformity* (Reading: Addison-Wesley, 1969).

may cause them to refrain from violations of the dictates of international institutions even in cases where violations may seem attractive in purely material terms.

Collective-Choice Considerations

Quite apart from these decision variables, the internal decision-making processes of the members of international society often play a role in producing compliance with the dictates of international regimes.[71] One such factor arises from the pervasive phenomenon of bureaucratic politics.[72] In essence, individual agencies within national governments sometimes come to define their roles, at least in part, in terms of administering and maintaining the provisions of one or more international regimes. This is, of course, obvious in the case of foreign ministries, which regularly become defenders of an array of institutional arrangements from arms-control regimes to regimes concerned with the conservation of wildlife or the protection of ecosystems.[73] Increasingly, however, other agencies have come to acquire similar stakes in the success of international regimes. In the United States, the National Marine Fisheries Service of the Department of Commerce has much to gain from the success of the international regime for whaling, and the Fish and Wildlife Service of the Department of the Interior has real stakes in the achievements of a number of international arrangements dealing with migratory birds as well as endangered species more generally. In such cases, individual agencies regularly find that their success in protecting agency turf and in competing for scarce resources depends, in part, on their ability to portray the regimes they work with as important and successful international arrangements. And this, in turn, generally hinges on an ability to demonstrate that the level of compliance with the provisions of such regimes is high. As a result, responsible agencies typically become staunch advocates for compliance with the terms of various international regimes in the bargaining processes that occur within governments as they move toward decisions on specific issues.[74]

71. For a discussion emphasizing linkages between international regimes and the domestic politics of the members see Stephan Haggard and Beth A. Simmons, "Theories of International Regimes," *International Organization* 41 (1987), 491–517.

72. For a well-known account of bureaucratic politics consult Graham T. Allison, *Essence of Decision* (Boston: Little, Brown, 1971), especially chap. 5.

73. For the details of a variety of international regimes dealing with wildlife see Lyster, *International Wildlife Law.*

74. For a seminal discussion of the analysis of governments as arenas for collective decision making rather than as unitary actors see Anthony Downs, *An Economic Theory of Democracy* (New York: Harper and Row, 1957), chap. 15.

International regimes also commonly give rise to nongovernmental interest groups committed to defending the provisions of specific regimes and prepared to press governments to comply with their dictates. In fact, the establishment of a regime can stimulate the growth of powerful interest groups in a number of the member states, which then form transnational alliances in order to persuade responsible agencies to comply with the requirements of the regime. To illustrate, the transnational network of marine scientists and ecologists that has emerged in conjunction with the environmental regime for the Mediterranean Sea appears to have become an important factor in encouraging governments to comply with the terms of the Mediterranean Action Plan and in triggering a storm of adverse publicity when governments fail to comply or seek to back down from their commitments under the plan.[75] A similar story emerges from an examination of the watchdog role of a wide range of nongovernmental organizations that push governments to comply with the terms of such diverse regimes as the water-quality arrangements set forth in the Great Lakes Water Quality Agreements of 1972 and 1978, the arrangements for endangered species incorporated in the Convention on International Trade in Endangered Species of Fauna and Flora, or the arrangements for Antarctica articulated in the Antarctic Treaty of 1959.[76] Conversely, there are good reasons to doubt whether responsible agencies will maintain favorable records of compliance with the requirements of international regimes in the absence of continuous pressure from nongovernmental organizations to do so. The comparatively underdeveloped nature of such pressures in the case of deep-seabed mining, in fact, constitutes one of the major reasons why many observers are skeptical about the prospects of success for the deep-seabed mining regime articulated in Part XI of the Law of the Sea Convention of 1982.[77]

In some cases, moreover, compliance with the requirements of international regimes becomes routinized through inclusion in the standard operating procedures of responsible agencies.[78] As many observers

75. Peter Haas, "Do Regimes Matter? A Study of Evolving Pollution Control Policies for the Mediterranean Sea," paper presented at the 1987 annual meetings of the International Studies Association.

76. On the case of the Great Lakes see the essays in *Alternatives* 13 (August/September 1986), a special issue entitled "Saving the Great Lakes." On CITES, consult Lyster, *International Wildlife Law*, chap. 12.

77. On the sources of this situation, with special reference to the United States, see Eckert, *Enclosure of Ocean Resources*, especially chap. 8, and James L. Malone, "Who Needs the Sea Treaty?" *Foreign Policy* 54 (1984), 44–63.

78. For a general account of standard operating procedures consult Allison, *Essence of Decision*, especially chap. 3.

have noted, agencies are ordinarily confronted with such a large volume of issues that they cannot begin to deal with the actions expected of them on a case-by-case basis. Accordingly, they must devise highly simplified decision rules to avoid becoming overwhelmed by the volume of their business. In fact, most agencies deal with the great majority of the issues that come their way through the routine application of such decision rules in the form of standard operating procedures. It follows that agencies charged with handling national responses to the requirements of international regimes are likely to comply with these requirements on a routine basis to the extent that compliance acquires the status of a standard operating procedure. There is a good deal of evidence to suggest that this basis of compliance has come to play an important role in ensuring high levels of compliance even in cases of highly politicized international regimes such as those involving arms-control arrangements.[79] It should come as no surprise, therefore, that compliance often becomes almost entirely routinized in connection with institutional arrangements that are less sensitive in political terms, like the regime governing the use of the electromagnetic spectrum or the regime dealing with marine living resources in the Antarctic region. Of course, this basis of compliance only becomes more compelling when responsible agencies come to rely on a metanorm or higher-order standard operating procedure that calls on decision-makers to use specific standard operating procedures routinely in handling their workload.[80]

What can we conclude from this discussion of the bases of compliance in international society? Briefly, the absence of authorized enforcement mechanisms does not justify the conclusion that the members of international society can simply ignore institutional dictates whenever they seem inconvenient or troublesome. In fact, we can identify a number of distinct bases of compliance at the international level, and there are good reasons to expect these considerations to exert effective pressure on most members of international society to comply with the dictates of institutional arrangements most of the time. Without doubt, violations of the rights and rules of international institutions can and will occur from time to time. But there is nothing remarkable about this. The occurrence of similar violations in domestic society does not cast doubt on our assumption that institutional arrangements are significant determinants of collective outcomes at the domestic level. The fact that

---

79. See Abram Chayes, "An Enquiry into the Workings of Arms Control Agreements," *Harvard Law Review* 85 (1972), 905–969, and Fisher, *Improving Compliance*, chap. 7.

80. For a suggestive discussion of metanorms see Robert Axelrod, "An Evolutionary Approach to Norms," *American Political Science Review* 80 (1986), 1100–1102.

violations occur from time to time is not incompatible with the proposition that the dictates of institutional arrangements operate as major constraints on the behavior of individual actors in social systems. And there is no evidence to suggest that compliance levels are, for the most part, lower in international society than they are in domestic societies.

Conclusion

The implications of the contrasts between domestic society and international society outlined in the first section of this chapter are often exaggerated, at least when it comes to assessing the significance of institutional arrangements as determinants of collective outcomes. In the typical case, the members of international society cannot expect to be able to discard existing institutions whenever they seem inconvenient or troublesome and they cannot afford to violate the dictates of these arrangements on a casual basis. Of course, the character of existing institutions does not explain everything that occurs in international society. Nor are institutions actors in their own right, capable of directing the behavior of the members of international society on an intentional basis. But, surely, no one would think to advance such arguments to suggest that institutional arrangements are not significant determinants of collective outcomes in domestic society. The essential point is that social institutions, at the international level as well as at the domestic level, constitute major constraints on the behavior of individual actors and, as a result, determine a substantial portion of the variance in collective outcomes within all social systems. It follows that those who devote a good deal of their energy to the defense of institutions they find attractive or to the reform of institutions they regard as undesirable are not wasting their time. They may well underestimate the difficulty of the task of controlling or directing the course of institutional change in complex social systems. If institutions were easier to restructure, after all, they would be less significant as determinants of collective outcomes in human social systems. Undoubtedly, however, those who realize that alterations in institutional arrangements are apt to cause substantial changes in collective outcomes are on the right track, whether they are operating in domestic societies or in international society.

Chapter Four

Regime Dynamics: The Rise and Fall of International Regimes

Because international regimes are complex social institutions, it is tempting to approach them in static terms, abstracting away the effects of time and social change in the interest of getting a clearer fix on the interactions among the constituent elements of specific institutions. This practice, drastically simplifying the analysis of regimes, is certainly justifiable in some contexts. It makes sense, for instance, for those endeavoring to assess the probable consequences of specific regimes or to choose from a menu of alternative institutional arrangements. But this orientation is not sufficient to serve as a basis for a comprehensive analysis of international regimes. Like all social institutions, regimes change over time. As a result, we must examine the developmental patterns or life cycles of regimes. How can we account for the formation of any given regime? What factors determine whether an existing regime will remain operative over time? Can we shed light on the rise of new regimes by analyzing the decline of their predecessors? Are there discernible patterns in these dynamic processes? Is it possible to formulate nontrivial generalizations about the dynamics of international institutions?

REGIMES AS HUMAN ARTIFACTS

A distinctive feature of all social institutions, including international regimes, is the conjunction of behavioral regularities and convergent expectations.[1] This is not to suggest that both these elements must

1. See also A. Irving Hallowell, "The Nature and Function of Property as a Social Institution," *Journal of Legal and Political Sociology* 1 (1943), 115–138.

crystallize simultaneously for a regime to come into existence: the rise of behavioral regularities sometimes leads to a convergence of expectations and vice versa. Undoubtedly, mutual reinforcement between these elements also plays a role in the development and maintenance of many social institutions. The existence of such a conjunction, however, commonly produces identifiable social conventions, which actors conform to without making elaborate calculations on a case-by-case basis.[2] It follows that international regimes, like other social institutions, typically acquire a life of their own in the form of clusters of widely accepted social conventions.

This perspective makes it clear that regimes are human artifacts, which possess no existence or meaning apart from the activities of human beings (whether as individuals or as members of collective entities). In this sense, they belong to the sphere of social systems in contrast to natural systems. This hardly means that regimes will be easy to construct or simple to reform on the basis of deliberate planning or social engineering. It does, however, have several other implications. International regimes do not exist as ideals or essences prior to their emergence as outgrowths of patterned human behavior. It is therefore misleading to think in terms of discovering regimes.[3] Similarly, there is no such thing as an unnatural regime; they are all responses to collective-action problems among groups of human beings and products of the emergence of patterns in the behavior of individuals or collective entities. But this is not to say that it is irrelevant or uninteresting to assess the performance of specific regimes in terms of well-defined criteria of evaluation or to strive for the articulation of more appropriate regimes in concrete situations. Just as alternative language systems may yield more or less desirable results in terms of criteria such as precision of communication or richness of description, international regimes can have a substantial impact on the achievement of allocative efficiency, equity, ecological balance, and so forth. Accordingly, it makes perfectly good sense to endeavor to modify existing regimes in the interests of promoting efficiency, equity, or any other desired outcome.

Note, however, that international regimes, like other social institutions, are ordinarily products of the behavior of groups of actors

2. For a suggestive account of the nature and role of social conventions see Russell Hardin, "The Emergence of Norms," *Ethics* 90 (1980), 575–587, and *Collective Action* (Baltimore: Johns Hopkins University Press, 1982), chaps. 11–14.
3. This view has much in common with the philosophical tenets of legal positivism as contrasted with natural law. See the well-known exchange between H. L. A. Hart, "Positivism and the Separation of Law and Morals," *Harvard Law Review* 71 (1958), 593–629, and Lon L. Fuller, "Positivism and Fidelity to Law: A Reply to Professor Hart," *Harvard Law Review* 71 (1958), 630–671.

(whether individuals or collective entities). While any given regime reflects the behavior of all those participating in it, individual actors who operate in an uncoordinated fashion are seldom able to exercise much influence over the character of the regime.[4] This does not mean that regimes, treated as complex social institutions, never undergo rapid changes or transformations. As the collapse of the institution of colonialism in the middle of the twentieth century and the recent disintegration of the traditional system governing various uses of the oceans demonstrate, change can spread rapidly; the linkages among the elements of institutions may actually accelerate processes of change once they get started. Nonetheless, it is far from easy to bring about planned or guided changes in social institutions. Behavioral regularities and convergent expectations frequently prove resistant to change, even when they produce outcomes that are widely understood to be suboptimal or undesirable for all concerned. Existing institutional arrangements, such as the regimes governing Antarctica or the international coffee trade, are familiar constructs, whereas new arrangements require actors to assimiliate alternative patterns of behavior and to accept (initially) uncertain outcomes. Additionally, planned changes in regimes require not only a dismantling of existing institutions but also the coordination of expectations around new focal points.[5] Given the extent and the severity of conflicts of interest in international society, it is fair to assume that the convergence of expectations around new institutional arrangements will often be slow in coming. This problem is well-known in connection with legislative conferences in international society (consider the United Nations Conference on the Law of the Sea as a case in point). And it is apt to prove even more severe with respect to coordination arising spontaneously from the day-to-day behavior of individual actors expected to become subjects of any new or modified regime.

What is more, social institutions commonly encompass an array of informal as well as formal elements. Deliberate efforts to modify or reform international regimes, which ordinarily focus on the formal elements of these institutional arrangements, can therefore easily produce disruptive consequences neither foreseen nor intended by those promoting specific changes. Under the circumstances, there is generally some risk that ventures in the social engineering of international regimes will ultimately do more harm than good. The desire to engage in

4. This observation is of course a cornerstone of the analysis of competitive markets in microeconomics. For a clear exposition stressing this point see Francis M. Bator, "The Simple Analytics of Welfare Maximization," *American Economic Review* 47 (1957), 22–59.
5. On the role of focal points in the convergence of expectations see Thomas C. Schelling, *The Strategy of Conflict* (Cambridge: Harvard University Press, 1960), chap. 4.

social engineering in this realm is understandably strong, and I certainly do not mean to suggest that all such efforts are doomed to failure.[6] Also, situations sometimes arise (for example, as a result of the collapse of some preexisting regime) in which it is difficult to avoid conscious efforts to create or reform specific institutional arrangements. The situation facing those concerned with international monetary affairs in the aftermath of the Great Depression and World War II is a clear-cut case in point. But these comments do suggest that naive expectations regarding the efficacy of social engineering will constitute a serious shortcoming among policymakers and students of international relations alike.

REGIME FORMATION

What can we say about the origins of international regimes or the developmental processes through which regimes come into existence? In the Introduction to this book, I argued that all social institutions are responses to collective-action problems or situations in which the pursuit of interests defined in purely individualistic terms regularly leads to socially undesirable outcomes. As the literature on the prisoner's dilemma, the tragedy of the commons, the security dilemma, and other well-known collective-action problems clearly indicates, difficulties of this sort are pervasive at all levels of human activity.[7] Among other things, this observation helps to account for the common emphasis on the normative character of social conventions and the widespread desire to socialize those subject to institutional arrangements to conform to the requirements of social practices as a matter of course. But it tells us little about the actual processes through which international regimes arise. Is there a uniform developmental sequence in regime formation or are there several distinct processes? Though we are not yet in a position to answer this question definitively, my work on regimes has led me to conclude that institutional arrangements arise in several different ways in international society.

Developmental Sequences

Some regimes are properly interpreted as *self-generating* or *spontaneous* arrangements. They are, as Hayek puts it, "the product of the action of

---

6. See also Geoffrey Brennan and James M. Buchanan, *The Reason of Rules: Constitutional Political Economy* (Cambridge: Cambridge University Press, 1985). And for an analysis of strategies available to individual actors seeking to promote changes in prevailing institutions consult Victor P. Goldberg, "Institutional Change and the Quasi-Invisible Hand," *Journal of Law and Economics* 17 (1974), 461–492.

7. See, among others, Hardin, *Collective Action;* Mancur Olson, Jr., *The Logic of Collec-*

many men but . . . not the result of human design."[8] Such regimes are distinguished by the facts that they do not involve conscious coordination among participants, do not require explicit consent on the part of subjects or prospective subjects, and are highly resistant to efforts at social engineering. Though the terms "self-generating" and "spontaneous" are attributable to Hayek, Lewis covers some of the same ground in his study of social conventions, Schelling evidently has a similar phenomenon in mind in his discussion of interactive behavior, and Axelrod points to analogous processes in the evolution of cooperation.[9] In fact, there are several realms in which the expectations of subjects regularly converge to a remarkable degree in the absence of conscious design or even consciousness. Natural markets are an important case in point well-known to social scientists, and similar processes are at work in many balance-of-power situations at the international level. Spontaneous arrangements relating to such things as language systems and social mores are even more striking in this regard. As those who have tried can attest, it is extraordinarily difficult to create an effective language by design. Yet large groups of individuals are perfectly capable of converging on relatively complex linguistic conventions and of using them proficiently without high levels of consciousness.

The processes through which spontaneous arrangements arise are not well-understood.[10] Surely, the propositions of sociobiology are not sufficient to provide a satisfactory account of the formation of institutional arrangements that take such diverse forms and change so rapidly. And social psychology offers no comprehensive theoretical account of interactive learning relevant to the emergence of social conventions.[11] There are intriguing hints concerning such processes, however, in recent work on the convergence of expectations and the evolution of cooperation. Thus, Schelling's account of tacit bargaining has highlighted the role of what he terms *focal points* or *salient solutions* in coordinating the behavior of interdependent actors in the absence of explicit communication.[12] Perhaps even more suggestive in this context is the

tive Action (Cambridge: Harvard University Press, 1965); and Garrett Hardin and John Baden, eds., *Managing the Commons* (San Francisco: W. H. Freeman, 1977).

8. Friedrich A. Hayek, *Rules and Order*, vol. 1 of *Law, Legislation, and Liberty* (Chicago: University of Chicago Press, 1973), 37.

9. David K. Lewis, *Convention: A Philosophical Study* (Cambridge: Harvard University Press, 1969); Thomas C. Schelling, *Micromotives and Macrobehavior* (New York: Norton, 1978); and Robert Axelrod, *The Evolution of Cooperation* (New York: Basic Books, 1984).

10. For some suggestive observations, however, see Hardin, *Collective Action*, chaps. 11–14.

11. But consider the research on the norm of reciprocity reviewed in Kenneth J. Gergen, *The Psychology of Behavior Exchange* (Reading: Addison-Wesley, 1969).

12. Schelling, *Strategy of Conflict*, especially chap. 4.

analysis Axelrod has developed to account for the emergence of norms in large groups.[13] In essence, Axelrod argues that individuals or individual actors engage in a learning process based on trial and error coupled with a form of natural selection.[14] Under such conditions, groups of actors can coordinate their actions around common conventions, or, in Axelrod's terms, norms, by discarding comparatively unsuccessful behaviors and retaining more successful ones through a sequence of trials. Though much of the resultant discussion actually focuses on reasons why actors are likely to comply with conventions or norms once they come into existence, Axelrod does provide an interesting account of the spread of behavioral prescriptions in response to the preferences of unusually influential actors as well as to the special importance of maintaining credible reputations in highly decentralized social systems.[15] Whatever the nature of the processes giving rise to such arrangements, it is easy enough to understand the attractions of self-generating or spontaneous social institutions. They are capable of contributing significantly to the welfare of the members of any given society in the absence of high transaction costs or formal restrictions on the liberties of individual participants.[16] What is more, they obviate the need to construct fictitious accounts of the negotiation or articulation of social contracts.

A strikingly different developmental sequence can be described under the rubric of *negotiated* institutional arrangements. These are regimes characterized by conscious efforts to agree on their major provisions, explicit consent on the part of individual participants, and formal expression of the results. At the outset, it is worth differentiating among several types of negotiated regimes in international society. Such regimes may take the form either of constitutional contracts or legislative bargains. In constitutional contracts (for example, the regime for Antarctica) those expecting to be subject to institutional arrangements participate directly in the relevant negotiations.[17] Legislative bargains (for example, the various United Nations efforts to devise a regime for Palestine), by contrast, occur under conditions in which those expected

13. Robert Axelrod, "An Evolutionary Approach to Norms," *American Political Science Review* 80 (1986), 1095–1111.

14. On learned behavior pertaining to institutional arrangements see also Ernst B. Haas, "Words Can Hurt You; or, Who Said What to Whom about Regimes," 23–59 in Stephen D. Krasner, ed., *International Regimes* (Ithaca: Cornell University Press, 1983).

15. Axelrod, "An Evolutionary Approach to Norms," 1008–1009.

16. But note that such institutions may rely on effective, if informal, social pressures. For a general discussion of social pressure consult C. A. Kiesler and Sara B. Kiesler, *Conformity* (Reading: Addison-Wesley, 1969).

17. For a general account of constitutional contracts see James M. Buchanan, *The Limits of Liberty* (Chicago: University of Chicago Press, 1975), especially chap. 4.

to be subject to a regime do not participate directly but are only represented in negotiations by others. It is useful, as well, to distinguish between comprehensive negotiated regimes and those that can be described as partial or piecemeal. Comprehensive regimes (for example, the emerging arrangements for the deep seabed) sometimes flow from careful and orderly negotiations. Given the conflicts of interest prevalent in international society, however, it is to be expected that negotiated arrangements will often exhibit a piecemeal quality, leaving many issues to be settled through practice and precedent.[18] Negotiated regimes are common at the international level. In fact, there is some tendency to become so caught up in the analysis of negotiated international regimes that it is easy to forget that other developmental sequences are also prominent.

Any effort to understand the formation of negotiated regimes requires a careful analysis of bargaining. The existing body of theoretical and empirical work on bargaining can be brought to bear on the study of regime dynamics.[19] For example, this sort of regime formation can be cast in terms of the theory of N-person, nonzero-sum, cooperative games or in terms of the microeconomic models that originated in the Edgeworth box construct and were developed by Zeuthen, Pen, and Cross.[20] Though this work is helpful, it also highlights some of the major gaps in our understanding of regime dynamics. Theoretical models of bargaining are well-known for their tendency to yield indeterminate or conflicting results, and much of the empirical work on bargaining emphasizes the importance of an array of contextual factors. Additionally, the analytic literature on bargaining tends to abstract away a number of factors that are important in connection with international regime formation (for example, incomplete information, unstable preferences). Among other things, this tendency has resulted in a lack of concern for factors that can lead to failure to reach agreement on the terms of negotiated regimes, even when there are feasible options that would yield gains for all parties. To illustrate, the disruptive potential of strategic moves, free riding, the absence of suitable compliance mechanisms,

18. This insight is, of course, examined extensively in the neofunctionalist literature on regional integration. See, for example, the essays in Leon Lindberg and Stuart Schcingold, eds., *Regional Integration: Theory and Practice* (Cambridge: Harvard University Press, 1971).

19. For a review of the major theories of bargaining see Oran R. Young, ed. and contributor, *Bargaining: Formal Theories of Negotiation* (Urbana: University of Illinois Press, 1975).

20. On the game-theoretic models consult R. Duncan Luce and Howard Raiffa, *Games and Decisions* (New York: Wiley, 1957). The microeconomic models are reviewed in Young, *Bargaining*, pt. 2.

and so forth is commonly overlooked or deemphasized in the theoretical literature on bargaining.[21]

A third category of international regimes entails *imposed* arrangements. Imposed arrangements differ from self-generating or spontaneous regimes in that they are fostered deliberately by dominant powers or consortia of dominant powers. At the same time, such regimes typically do not involve explicit consent on the part of subordinate actors, and they often operate effectively in the absence of any formal expression. In short, imposed regimes are established deliberately by dominant actors who succeed in getting others to conform to the requirements of these arrangements through some combination of coercion, cooptation, and the manipulation of incentives.[22] In the classic case, a hegemonic actor openly and explicitly articulates institutional arrangements and compels subordinate actors to conform to them. Classical feudal arrangements as well as many of the great imperial systems exemplify this pattern.[23]

As Kindelberger and others have pointed out, such overt hegemony is almost certainly the exception rather than the rule, even in international society. What is more common are various forms of leadership in which an actor (or a small group of actors) that is markedly superior to others in natural and human resources or other bases of power plays a critical role in designing institutional arrangements and inducing others to agree to their terms.[24] The leader may exercise influence by threatening others with negative outcomes (for example, termination of aid or withdrawal of trading privileges) or by offering rewards for cooperation (for example, access to advanced technology or loans on favorable terms). But leadership differs from overt hegemony in that it involves a distinct element of negotiation or give-and-take in contrast to processes in which an obviously dominant actor simply dictates terms to others who have no choice but to acquiese. In such cases, the distinction between imposed regimes and negotiated regimes begins to blur, and it is not helpful to insist on a hard-and-fast separation between the two. In effect, true hegemony constitutes an extreme case, while leadership encompasses a range of cases in which one party (or small group) possesses substantially

21. For a discussion of these problems in the context of international bargaining see Oran R. Young, *The Politics of Force: Bargaining during International Crises* (Princeton: Princeton University Press, 1968).

22. See also Robert Gilpin, *The Political Economy of International Relations* (Princeton: Princeton University Press, 1987).

23. A. P. Thornton, *Doctrines of Imperialism* (New York: Wiley, 1965).

24. Charles P. Kindleberger, "Dominance and Leadership in the International Economy," *International Studies Quarterly* 25 (1981), 242–254; "International Public Goods without International Government," *American Economic Review* 76 (1986), 1–13; and "Hierarchy versus Inertial Cooperation," *International Organization* 40 (1986), 841–847.

greater bargaining power than the others. This conception of leadership undoubtedly applies to the role of Great Britain in the formation of the nineteenth-century regime for the oceans as well as to the role of the United States in the development of the postwar international trade and monetary regimes.[25]

The dynamics of imposed regimes must be approached in terms of power, despite the well-known problems afflicting efforts to come to terms with the phenomenon of power.[26] Several observations about the exercise of power in connection with regime formation are worth emphasizing at the outset. There is no reason to assume that dominant actors must coerce subordinate actors continuously to ensure their conformity to the requirements of imposed institutional arrangements. Habits of obedience on the part of subordinate actors regularly arise over time.[27] Most forms of dependence have a strong cognitive component as well as some structural basis. And the recent literature on core-periphery relations makes it clear that the methods through which hegemonic powers acquire and maintain dominance in institutionalized relationships are often highly complex.[28] It should come as no surprise, therefore, that the most successful imposed regimes have not been characterized by continuous exercises of overt coercion.[29] There are, in addition, significant costs associated with the role of hegemon. Hegemons frequently find themselves paying off subordinate actors to remain docile, as in the case of the economic benefits the Soviet Union provides to other socialist states. Similarly, hegemonic actors generally bear the burden of responsibility for the performance of the regimes they impose, a fact that almost always forces them to forego positions of moral or ethical leadership in the social systems in which they operate.

The Route Taken

How can we explain which of these developmental sequences will govern the formation of specific international regimes? Why were serious efforts made to reach agreement on negotiated arrangements for

25. See also Gilpin, *Political Economy of International Relations*, and "The Politics of Transnational Economic Relations," *International Organization* 25 (1971), 398–419.

26. For a review of efforts to conceptualize power in international society consult David A. Baldwin, "Power Analysis and World Politics," *World Politics* 31 (1979), 161–194.

27. For an insightful discussion of the role of habits of obedience see H. L. A. Hart, *The Concept of Law* (Oxford: Oxford University Press, 1961), 49–64.

28. Consult Michael Hechter, *Internal Colonialism* (Berkeley: University of California Press, 1975).

29. For a richly illustrated account emphasizing the role of structural factors see Stephen D. Krasner, *Structural Conflict: The Third World against Global Liberalism* (Berkeley: University of California Press, 1985).

the oceans during the 1970s and 1980s when regimes for marine resources typically took the form of imposed or spontaneous arrangements in the past? Why have we come to rely increasingly on negotiated regimes for various commodities when spontaneous arrangements (for example, natural or unregulated markets) would have seemed perfectly adequate in the past? The first thing to notice in reflecting on these questions is that the three developmental sequences I have identified are not mutually exclusive under real-world conditions. An imposed or spontaneous regime, for example, is sometimes codified or legitimated in a formal, constitutional contract; the 1958 Geneva Convention on the Continental Shelf clearly illustrates this phenomenon. The promulgation of a negotiated regime will have little effect unless its precepts and requirements are absorbed into the routine behavior of the participants. Efforts to translate the terms of regional-fisheries arrangements into day-to-day management systems, for example, indicate clearly how difficult it can be to implement negotiated arrangements in international society.[30] By the same token, imposed regimes are sometimes increasingly accepted as legitimate with the passage of time, with the result that it becomes less necessary for dominant actors to coerce others into conforming with their requirements. A transition of this sort may well have occurred in recent years in the management authority of adjacent coastal states over marine fisheries. It follows that any attempt to classify actual international regimes rigidly in terms of the three developmental sequences is apt to distort reality and to produce confusion rather than enhance understanding.

Nonetheless, we are still faced with the problem of identifying the factors that lead to the predominance of one developmental sequence or another. There is, in my judgment, a pronounced tendency to exaggerate the role of negotiated regimes in contrast to imposed or spontaneous regimes at the international level. Such an emphasis on negotiated arrangements appeals to the conceptions of rational choice and conscious institutional design that pervade the contemporary literature on public policy and public choice. Additionally, a focus on spontaneous regimes seems to imply an organic conception of society, an orientation that is often associated with illiberal political views.[31] Yet it is hard to escape the conclusion that spontaneous regimes are of critical impor-

30. For an account of a number of regional fisheries arrangements see J. A. Gulland, *The Management of Marine Fisheries* (Seattle: University of Washington Press, 1974), especially chap. 7.

31. But note that some radical thinkers have also espoused organic conceptions of society. See, for example, Peter Kropotkin, *Mutual Aid: A Factor of Revolution* (New York: New York University Press, 1972).

tance in international society, just as they are in other social settings. Even in cases where a new institutional arrangement is articulated in a treaty or convention, formalization is often better understood as a codification of behavioral patterns that have arisen spontaneously than as the promulgation of a new institutional arrangement requiring dramatic changes in existing behavioral patterns. Many of the substantive provisions proposed for inclusion in the law-of-the-sea convention, for example, are properly understood as illustrations of this phenomenon. A number of those who have contributed to the emerging literature on the formation and maintenance of international institutions also argue that powerful leadership, if not overt hegemony, plays a role of great importance.[32]

Other things being equal, the incidence of negotiated institutional arrangements will vary with the degree of centralization of authority in a society. Negotiated regimes can be expected to be pervasive in societies in which the state is highly developed and not severely constrained in functional terms. This proposition would account for the lower incidence of negotiated arrangements in international society than in domestic societies as well as for the growing role of negotiated arrangements in advanced industrial societies.[33] At the same time, the prominence of imposed regimes will vary inversely with the level of interdependence in a society. The growth of interdependence increases the capacity of all relevant actors to inflict costs on each other, and this condition serves to blur (if not to eliminate) the distinction between dominant and subordinate actors.[34] This proposition would explain the higher incidence of imposed regimes in international society than in domestic societies, as well as in traditional societies in contrast to advanced industrial societies. Curiously, increases in the complexity of social systems can heighten the importance of spontaneous arrangements over that of imposed or negotiated arrangements. It is not surprising that the ability of dominant actors to impose institutional arrangements generally declines as a function of social complexity. But it is noteworthy that the capacity of groups of actors to arrive at meaningful or coherent bargains is apt to decline as the issues at stake become

32. For evaluations of this "hegemonic stability" perspective see Robert O. Keohane, *After Hegemony: Cooperation and Discord in the World Political Economy* (Princeton: Princeton University Press, 1984), and Duncan Snidal, "The Limits of Hegemonic Stability Theory," *International Organization* 39 (1985), 579–614.

33. For a sophisticated treatment of the distinguishing features of international society see Hedley Bull, *The Anarchical Society* (New York: Columbia University Press, 1977).

34. Oran R. Young, "Interdependencies in World Politics," *International Journal* 24 (1969), 726–750.

increasingly complex.[35] As a result, spontaneous arrangements arising from interactive behavior are prominent in modern societies, despite the fact that this phenomenon runs counter to the widespread propensity to regard such institutions as irrational or illiberal.[36] As well, expansions in the scale of social systems (measured in terms of the size of the membership) will ordinarily militate against reliance on negotiated regimes and favor spontaneous or imposed arrangements. In very large systems, it is hard for the members to play a meaningful role in the negotiation of institutional arrangements; even the idea of explicit consent eventually begins to lose significance.[37] Of course, it is possible to offset these problems somewhat through the development of systems of representation. But the success of any system of representation depends critically not only on the presence of well-informed constituents but also on the maintenance of high standards of accountability in relationships between representatives and their constituents. It should come as no surprise, therefore, that international regimes exhibiting the superficial appearance of negotiated arrangements are sometimes better understood as imposed regimes in de facto terms.

No doubt, there are other approaches to explaining the incidence of these developmental sequences. Perhaps some of the propositions associated with sociobiology can be brought to bear on this question.[38] Some observers will surely want to argue that there are cultural factors at work in this context.[39] Those familiar with the recent literature on public choice will have something to say about the difficulties of arriving at negotiated regimes in constitutional or legislative settings, especially those in which voting plays an important part.[40] However, I am convinced that structural factors of the sort emphasized in the preceding paragraph are of central importance in accounting for the incidence of different developmental sequences.[41]

35. See also Hayek, *Rules and Order*, chap. 2.

36. For an account stressing the significance of spontaneous institutional arrangements in contemporary societies see Schelling, *Micromotives and Macrobehavior*.

37. Put in other language, the transaction costs of reaching negotiated agreements rise rapidly as a function of group size. See also E. J. Mishan, "The Postwar Literature on Externalities: An Interpretive Essay," *Journal of Economic Literature* 9 (1967), 21–24.

38. The seminal work on sociobiology is Edward O. Wilson, *Sociobiology: The New Synthesis* (Cambridge: Harvard University Press, 1975). For a selection of critiques of the major propositions of sociobiology consult Arthur L. Caplan, ed., *The Sociobiology Debate* (New York: Harper and Row, 1978).

39. For a well-known account stressing the role of culture in international relations see Adda B. Bozeman, *Politics and Culture in International History* (Princeton: Princeton University Press, 1960).

40. For a survey of this literature see Norman Frohlich and Joe A. Oppenheimer, *Modern Political Economy* (Englewood Cliffs: Prentice-Hall, 1978), especially chap. 1.

41. Students of international relations have employed the terms *structure* and *structural-*

Does It Make a Difference?

In the light of this discussion, it seems important to ask whether it makes a difference if international institutions arise in the form of spontaneous regimes, negotiated regimes, or imposed regimes. Unless the answer to this question is affirmative, the distinctions I have been developing in this chapter might well be dismissed as points of no more than academic interest.

The obvious place to begin in thinking about this question is with a consideration of outcomes or consequences. Is one type of regime more likely than another to promote allocative efficiency, equity, peace, ecological integrity, and so forth in the governance of international activities? As it happens, this is a highly complex matter with respect to which we are not yet in a position to offer definitive answers. Interestingly, however, there is a good deal to be said for the virtues of spontaneous arrangements from this point of view.[42] Language systems that arise spontaneously, for example, produce extraordinary social benefits in a highly efficient fashion. Much the same can be said of unregulated markets, at least when certain conditions pertaining to information, competition, and social costs obtain. And spontaneous arrangements yield these results in the absence of high transaction costs. They do not give rise to oppressive procedural requirements or armies of officials charged with implementing and enforcing the terms of formalized regimes; participants need not even be consciously aware of their existence. Nor do spontaneous arrangements require elaborate formal restrictions on the liberty of individual actors, though they ordinarily do give rise to effective forms of social pressure. Negotiated regimes, by contrast, typically result in high transaction costs and the progressive introduction of more and more intrusive restrictions on individual liberty.[43] Moreover, the introduction of a negotiated regime can hardly be said to provide any assurance of the achievement of allocative efficiency. Imposed regimes are intended to benefit dominant actors. Otherwise, those occupying dominant positions would have no incentive to expend resources on the formation of regimes.[44] And domi-

---

ism in a variety of ways. My emphasis here is on the idea that social systems have properties, such as centralization, interdependence, or complexity, that are attributes of the systems per se rather than of their constituent members. For further discussion see Kenneth N. Waltz, *Theory of International Relations* (Reading: Addison-Wesley, 1979).

42. For a suggestive, though overly optimistic, account of these virtues see Hayek, *Rules and Order,* chap. 2.

43. Though the neoconservative movement has recently stressed this point, it is worth noting that it has long been a major theme in the literature of anarchism. See, for example, Daniel Guerin, *Anarchism: From Theory to Practice* (New York: Monthly Review Press, 1970).

44. See also Gilpin, *Political Economy of International Relations.*

nant actors are often oriented more toward rent-seeking behavior than toward the pursuit of allocative efficiency—as the history of mercantilism attests. What is more, imposed regimes may become expensive to maintain, unless the hegemon or leadership group succeeds in inducing subordinate actors to accept the arrangements as legitimate.[45]

Of course, it is true that spontaneous regimes may yield outcomes that are difficult to justify in terms of any reputable standard of equity. Unregulated markets certainly exemplify this observation under a wide range of conditions. Unfortunately, however, negotiated regimes and, especially, imposed regimes cannot be counted on to yield outcomes that are more attractive on this account. This proposition is obviously true of imposed arrangements, which are designed to advance the interests of one or a few dominant actors. But it is noteworthy that negotiated regimes offer no guarantee of outcomes that are more defensible in terms of equity. Bargains struck may be heavily affected by an unequal distribution of bargaining strength, and it is often hard to foresee the consequences likely to flow from complex institutional arrangements in any case. Even if a negotiated regime is equitable in principle, moreover, there is usually considerable scope for implementing it in ways that differ greatly with respect to equity.[46]

By contrast, the situation strikes me as markedly different when we turn from the question of outcomes to a consideration of the stability of international regimes or the capacity of institutional arrangements to adjust to a changing environment in an orderly fashion.[47] It is here that spontaneous regimes often run into more or less severe problems. As the cases of language systems and moral codes suggest, these arrangements are particularly well suited to relatively settled social environments. The convergence of expectations through spontaneous processes takes time, especially in situations where a multiplicity of opinion leaders can be expected to direct attention toward divergent focal points concerning behavioral standards. Rapid social change, therefore, typically erodes spontaneous regimes without creating conditions conducive to the formation of new arrangements.

Imposed regimes and negotiated regimes, on the other hand, ordinarily stand up better in the face of social change. A flexible hegemon

45. For a rich account of the erosion of the legitimacy of British imperialism see A. P. Thornton, *The Imperial Idea and Its Enemies: A Study in British Power* (Garden City: Anchor Books, 1968).

46. This problem is largely abstracted away by writers like Rawls who assume perfect compliance with the principles of justice accepted by actors in the original position. See John Rawls, *A Theory of Justice* (Cambridge: Harvard University Press, 1971), 351.

47. See also Oran R. Young, "On the Performance of the International Polity," *British Journal of International Studies* 4 (1978), 191–208.

or leadership group can succeed in manipulating incentives or adjusting the terms of an imposed arrangement substantially, so long as its own position of dominance is not severely eroded by social change. Some students of international relations have been so struck by these considerations, in fact, that they have developed a conception of hegemonic stability in which the presence of a dominant actor or group figures as a necessary condition for the maintenance of institutional arrangements.[48] Parties to negotiated regimes can modify or revise the relevant arrangements on a deliberate basis in response to social change. In the case of constitutional contracts, for example, there is ordinarily nothing to prevent the members from amending or even replacing major provisions of an existing regime. The ongoing negotiations aimed at adapting the Antarctic regime to cover the possibility of exploiting commercially valuable minerals exemplify this prospect.[49] It is true, of course, that such responses will sometimes lead to incoherence in international regimes, since additions to existing arrangements are not always easy to square with the original character of a regime. Still, it is easy enough to see the attractions of negotiated regimes during periods of rapid social change.

All this poses a dilemma of sorts. Negotiated regimes and even imposed regimes are attractive in periods, like the present, characterized by rapid social change. Also, negotiated arrangements certainly appeal to those convinced of the efficacy of social engineering. Yet spontaneous regimes are not without real attractions in terms of the outcomes they are likely to produce. Under the circumstances, we find ourselves today in an era that features a growing emphasis on negotiated regimes at the international level, though we have only a limited understanding of how to manage such arrangements in a cheap and efficient way, much less in a fashion that will ensure equitable outcomes. It follows that we need to think much more systematically about the extent to which the problems of negotiated regimes are endemic or, alternatively, subject to alleviation through the development of suitable management techniques.

REGIME TRANSFORMATION

International regimes do not become static constructs even after they are fully articulated. Rather, they evolve continuously in response to their own inner dynamics as well as to changes in the political, economic,

48. See Keohane, *After Hegemony*, for an extended assessment of this argument.
49. M. J. Peterson, "Antarctica: The Last Great Land Rush on Earth," *International Organization* 34 (1980), 377–403.

and social environments. In this connection, I use the term *transformation* to refer to significant alterations in a regime's structure of rights and rules, the character of its social-choice procedures, and the nature of its compliance mechanisms. How extensive must these alterations be to produce qualitative change in the sense that we would want to speak of one regime disappearing and another taking its place? Does a shift from unrestricted common property to a regime of restricted common property for the marine fisheries, for example, entail a qualitative change of regimes? Answers to these questions must ultimately be arbitrary, and I shall not attempt to specify any generic threshold of transformation for international regimes.[50] Instead, I propose to focus on major alterations in existing regimes and to comment on the patterns of change giving rise to these alterations.

Patterns of Change

It is possible to identify several distinct processes leading to regime transformation. These processes sometimes revolve around factors that are endogenous to specific institutional arrangements. Some regimes harbor internal contradictions that eventually lead to serious failures and mounting pressure for major alterations. Such contradictions may take the form of irreconcilable conflicts among the constituent elements of a regime. For example, a regime that guarantees all participants unrestricted access to an area's resources while at the same time granting sovereignty over the area to a single member (like the regime articulated in the Svalbard treaty of 1920) is bound to generate serious frictions.[51] On the other hand, internal contradictions sometimes exhibit a developmental character, deepening over time as a consequence of the normal operation of a regime. Of course, this is the perspective underlying Marxist analyses of the capitalist international economic order.[52] At a more mundane level, however, the same sort of dynamic occurs in regimes governing many specific activities.[53] To illustrate, it is easy to trace the impact of evolving contradictions in unrestricted common-

50. Compare the well-known query posed by philosophers: How many Chevrolet parts added to a Ford automobile would it take to transform the vehicle from a Ford into a Chevrolet?

51. For a more general discussion of the problems posed by conflicts among rights consult Ronald Dworkin, *Taking Rights Seriously* (Cambridge: Harvard University Press, 1977), especially chap. 4.

52. For an illustration, see Immanuel M. Wallerstein, *The Capitalist World Economy* (New York: Cambridge University Press, 1979).

53. See the essays included in William R. Thompson, ed., *Contending Approaches to World System Analysis* (Beverly Hills: Sage Publications, 1983).

property regimes for marine fisheries during periods in which increasing harvest capacities lead to severe depletions of fish stocks.[54]

Several approaches to the analysis of these internal contradictions seem worth pursuing. It is relatively straightforward to conceptualize such phenomena in terms of the stability conditions associated with equilibrium models. That is, treating any given regime as a system of action, we can ask how far its central elements can be displaced before the system blows up rather than moves back toward a point of equilibrium.[55] Perhaps the best known example of this approach at the international level involves the reaction-process models devised by Richardson for the analysis of arms races.[56] Alternatively, it may prove helpful to examine internal contradictions leading to regime transformation in terms of the holistic perspective associated with dialectical reasoning.[57] This approach need not take the form of dialectical materialism or of any particular variety of Marxism; rather, its hallmarks are the analysis of complex institutions as dynamic wholes and a search for dialectical laws describing patterns of change in these entities.[58] It is worth pointing out that each of these approaches directs attention to the role of crises in regime transformation, whether such crises are characterized in terms of the vocabulary of systems going unstable or of some sort of collapse of an old order. We are now becoming familiar with discussions of liquidity crises in monetary regimes, crises of common property in regimes for renewable resources, and pollution crises brought on by such practices as the use of the air mantle as a sink for the disposal of residuals or wastes.[59] It follows that the recent literature exploring social applications of catastrophe theory is a promising source of insights into patterns of regime transformation.[60]

An alternative type of process leading to regime transformation places

54. For a seminal analysis consult H. Scott Gordon, "The Economic Theory of a Common Property Resource: The Fishery," *Journal of Political Economy* 62 (1954), 124–142.

55. For an accessible discussion of the stability conditions associated with equilibrium models see Anatol Rapoport, *Fights, Games, and Debates* (Ann Arbor: University of Michigan Press, 1960), pt. 1.

56. See also the more general account of reaction-process models in Kenneth Boulding, *Conflict and Defense* (New York: Harper and Row, 1962), especially chap. 2.

57. On the nature of dialectical reasoning see John Mepham and David H. Rubin, eds., *Issues in Marxist Philosophy*, vol. 1 (Atlantic Highlands, N.J.: Humanities Press, 1979).

58. For a thoughtful discussion of the nature of dialectical laws see Bertell Ollman, *Alienation: Marx's Theory of Man in Capitalist Society*, 2d ed. (New York: Cambridge University Press, 1976).

59. See also Richard A. Falk, *This Endangered Planet* (New York: Random House, 1971).

60. See also Michael B. Nicholson, *Formal Theories of International Relations* (Cambridge: Cambridge University Press, 1989), chap. 8.

greater emphasis on factors that are exogenous to specific institutional arrangements. Undoubtedly, shifts in the general structure of power in international society are often important in this connection. It is perhaps obvious that imposed regimes, especially those associated with true hegemony in contrast to more complex forms of leadership, are unlikely to survive for long in the wake of severe declines in the effective power of the dominant actor.[61] This may well help to account for the erosion in the postwar international monetary regime that has occurred during the 1970s and 1980s.[62] But it is important to recognize that both negotiated regimes and spontaneous regimes also reflect the prevailing structure of power in society. Institutional arrangements are never neutral in their impact on the interests of their members. Accordingly, powerful actors exert pressure whenever they can to devise constitutional contracts or legislative bargains favoring their interests.[63] And opinion leaders or pacesetters act in such a way as to move spontaneous arrangements into line with their own interests. Understandably, therefore, shifts in the distribution of power will be reflected, sometimes gradually rather than abruptly, in commensurate alterations in international regimes. In some instances, these changes are quite direct, involving power shifts in the immediate issue area covered by a given regime. There can be no doubt, for example, that recent changes in the regime set forth in the International North Pacific Fisheries Convention emerged directly from the expanding influence of the United States over the marine fisheries of the region. In other cases, the process is less direct in that the character of an international regime changes in response to broader shifts in the structure of power in international society as a whole. Thus, it is difficult to comprehend many features of recent efforts to restructure regimes for ocean resources without a sophisticated appreciation of the broader patterns of change in the distribution of power in international society that have unfolded in the recent past.[64]

The analysis of such patterns of regime transformation is hampered by both empirical and conceptual problems. The principal empirical limitation arises from the fact that we continue to lack a satisfactory measure of power, despite numerous efforts to devise a usable metric or index in this area.[65] As a result, while it is easy enough to recognize

61. Reinhold Niebuhr, *The Structure of Nations and Empires* (New York: Scribner's, 1959).

62. Gilpin, *Political Economy of International Relations*, chap. 4.

63. In short, regimes are seldom developed under conditions approximating the Rawlsian "veil of ignorance." See Rawls, *A Theory of Justice*, for an account of these conditions.

64. See also Joseph S. Nye, Jr., "Ocean Rule-making from a World Perspective," 221–244 in Ocean Policy Project, *Perspectives on Ocean Policy* (Washington, D.C., 1974).

65. For a critical review see David A. Baldwin, "Money and Power," *Journal of Politics* 33 (1971), 578–614.

major shifts in power after the fact, it is exceedingly difficult to pinpoint the early phases of significant shifts or to monitor such shifts closely as they unfold. For example, there is no doubt that the ability of the United States to control the international monetary regime has declined in recent years, but it is hard to say just how rapidly this trend is progressing and where it will lead during the near future.[66] On the conceptual level, the central problem arises from a lack of consensus on the definition of power. This problem is partly attributable to the complex and elusive character of the phenomenon of power.[67] In part, the problem arises from the fact that power plays a significantly different role in various analytic perspectives in common use among social scientists. Compare, for example, the conceptions of power embedded in the propositions of those who think in terms of structural bases of dependence and of those who focus on the behavior of individual actors and employ the language of interdependence.[68] It follows that this type of regime transformation is not well-understood at present. But this limitation is hardly a sufficient reason to deemphasize the role that shifts in the structure of power play in producing major alterations in specific regimes. Rather, I would agree with those who argue that this situation calls for a renewed effort to come to terms in a systematic fashion with power and shifts in the distribution of power.

Beyond this, however, international regimes regularly fall victim to other exogenous factors. Perhaps the most dramatic examples of such processes of transformation occur in conjunction with changes in the nature and distribution of technology.[69] To illustrate, the advent of large stern trawlers and factory ships after World War II decisively undermined many unrestricted common-property regimes governing marine fisheries, which had performed at least tolerably for a long time. Similarly, the rapid growth of satellite-communications technology is threatening to swamp the existing institutional arrangements that govern the use of the global electromagnetic spectrum.[70] But other ex-

66. For an account that even raises questions about the decline of American influence see Susan Strange, "The Persistent Myth of Lost Hegemony," *International Organization* 41 (1987), 551–574.

67. See also Baldwin, "Power Analysis."

68. To illustrate, compare the ideas set forth in Johan Galtung, "A Structural Theory of Imperialism," *Journal of Peace Research* 2 (1971), 81–118, with those outlined in Robert O. Keohane and Joseph S. Nye, Jr., *Power and Interdependence* (Boston: Little, Brown, 1977).

69. For a sweeping view of Western history that stresses the role of technological change see William H. McNeil, *The Rise of the West* (Chicago: University of Chicago Press, 1963). For a more specific argument relating technological change to many contemporary environmental problems see Barry Commoner, *The Closing Circle* (New York: Knopf, 1971).

70. See also Seyom Brown, Nina W. Cornell, Larry Fabian, and Edith Brown Weiss, *Regimes for the Ocean, Outer Space, and Weather* (Washington, D.C.: Brookings, 1977),

ogenous factors can produce equally far-reaching results in the transformation of international regimes. Major shifts in domestic priorities can make certain types of regimes unworkable at the international level. There can be no doubt, for instance, that growing attachments to policies aimed at full employment and social welfare at the domestic level have played a significant role in rendering fixed exchange rates unworkable in the international monetary regime. Much the same is true of major increases in demand for certain renewable resources arising from overall population growth or sharp shifts in tastes among consumers. Such developments surely constitute a major factor in recent problems with common-property arrangements for the marine fisheries or for pollution control in areas such as the Mediterranean Basin. Significant changes in one international regime can also trigger pressures for change in other institutional arrangements. To illustrate, any success in efforts to modify the existing regime for whaling that have the effect of encouraging the growth of stocks of great whales will have important implications for arrangements governing the harvest of renewable resources, such as krill, in the Southern Ocean.[71]

The impact of these exogenous factors is certainly difficult to predict in specific cases. The course of technological development is discontinuous and hard to foresee.[72] The processes through which human values and tastes evolve are poorly understood. It is even difficult to make accurate predictions of demographic shifts, despite the availability of empirical projections based on past trends. And, of course, a clear understanding of the interactions among changes that affect collections of international regimes presupposes the growth of knowledge of the whole issue of regime transformation. Nonetheless, it seems important to recognize the significance of exogenous factors in thinking about regime transformation. If nothing else, this recognition reminds us of the dangers of approaching specific institutional arrangements in isolation from the broader social setting.

Paths to Transformation

Which of these patterns of change will occur most frequently in regime transformation at the international level? How can we account

---

especially chaps. 11–13, and Gregory C. Staples, "The New World Satellite Order: A Report from Geneva," *American Journal of International Law* 80 (1986), 699–720.

71. On the natural resources of the Southern Ocean see G. L. Kesteven, "The Southern Ocean," 467–499 in Elisabeth Mann Borgese and Norton Ginsburg, eds., *Ocean Yearbook, 1* (Chicago: University of Chicago Press, 1978).

72. See also Lester B. Lave, *Technological Change: Its Conception and Measurement* (Englewood Cliffs: Prentice-Hall, 1966).

for differences in the incidence of various processes of transformation? Once again, it is helpful to begin with the observation that the patterns of change are not mutually exclusive; several of them can occur simultaneously, interacting with each other to produce a complex dynamic. There is little doubt, for example, that technological developments severely exacerbated internal contradictions built into the traditional regimes of unrestricted common property in the marine fisheries. And emerging contradictions in common-property arrangements that govern the disposal of various wastes or residuals have provided a stimulus for the development of radically different technologies in this area, a shift that is certain to generate pressures for major changes in pollution-control regimes. It follows that sophisticated analyses of the transformation of specific international regimes will typically require examination of several distinct processes of transformation together with the interactions among them. For many purposes, therefore, it will not help much to concern ourselves with efforts to determine the relative importance or weight of individual processes of transformation.

It is also worth noting that views regarding the importance of different processes of transformation commonly correlate closely with broader philosophical or ideological perspectives. Marxists and others who think in dialectical terms, for example, can be expected to approach the problem of regime transformation primarily in terms of the impact of internal contradictions.[73] They will search for dialectical laws governing regime dynamics and emphasize the growth of antithetical forces leading to the collapse or breakdown of existing institutional arrangements. Those whose thinking reflects geopolitical ideas, mercantilism, or realism, by contrasts, typically focus on structures of power and attribute significant changes in international regimes to broader shifts in the distribution of capabilities at the international level. They regularly treat existing institutional arrangements as reflections of the structure of international society as a whole, expecting specific regimes to change in predictable ways in the wake of shifts in the broader structure of power. Yet another perspective is characteristic of the thinking of many liberals who emphasize rational behavior and the mutual benefits of cooperation rather than dialectical processes or the central role of power.[74] They are inclined to interpret the transformation of regimes as attempts to achieve reasoned adjustments to exogenous developments such as tech-

73. For non-Marxian analyses of this type see G. W. F. Hegel, *The Philosophy of History*, trans. by J. Sibree (New York: Dover, 1956), and Oswald Spengler, *The Decline of the West*, abr. trans. by Charles F. Atkinson (New York: Knopf, 1962).

74. For a thoughtful account of this type see Ernst B. Haas, "Why Collaborate? Issue Linkage and International Regimes," *World Politics* 32 (1980), 357–405.

nological change or population growth.[75] Not surprisingly, they generally prefer to think in terms of the articulation of negotiated regimes, and they are among those most likely to exaggerate the scope for successful social engineering in international society.

Efforts to determine which of these general orientations is correct seldom yield illuminating results. Not only do the points of view approach the problem of regime transformation from radically different directions, they also rest on incompatible premises, which are difficult to test in any meaningful way.[76] Still, it is useful to bear these divergent perspectives in mind in exploring the question of regime transformation. Doing so is likely to increase the sophistication of our efforts to understand specific cases of regime transformation as well as to improve communication among those who share an interest in the problem of explaining the transformation of institutional arrangements.

It is tempting also to argue that spontaneous regimes, negotiated regimes, and imposed regimes typically differ with respect to the processes of transformation they undergo. At first glance, it seems reasonable to expect spontaneous regimes arising in the absence of human design to be more prone to internal contradiction than negotiated regimes, which arise in the form of conscious agreements. Similarly, imposed regimes, tied closely to the structure of power in international society, would appear to be more sensitive to shifts in the distribution of capabilities than spontaneous or negotiated arrangements. Yet this line of reasoning has serious flaws that are readily apparent on reflection. Major contradictions or elements of incoherence, for example, are common in constitutional contracts, which are typically products of political compromise rather than rational planning.[77] Though it is certainly true that imposed regimes are sensitive to shifts in the distribution of power, much the same is true of negotiated regimes and even spontaneous regimes. What is more common, for instance, than efforts to reinterpret constitutional contracts more or less drastically in the wake of significant shifts in the structure of power in a social system?[78] These observations

75. See Falk, *This Endangered Planet,* for an analysis reflecting this point of view.

76. For a general treatment of these issues see Thomas S. Kuhn, *The Structure of Scientific Revolutions,* 2d ed. (Chicago: University of Chicago Press, 1970). For an analysis of similar issues with particular reference to international politics consult Graham Allison, *The Essence of Decision* (Boston: Little, Brown, 1971).

77. See the observations on this phenomenon developed in Wolff's critique of Rawls: Robert Paul Wolff, *Understanding Rawls* (Princeton: Princeton University Press, 1977).

78. For an account that stresses the view that constitutional contracts are, in any case, little more than interpretations arising from a flow of authoritative decisions see Myres S. McDougal and Associates, *Studies in World Public Order* (New Haven: Yale University Press, 1960).

do not lead me to rule out the development of nontrivial generalizations about regime transformation along these lines. But it does seem clear that such generalizations must await the formulation of more subtle distinctions among the developmental sequences associated with regime formation that I introduced earlier in this chapter.

CONCLUSION

International regimes are human artifacts characterized by a conjunction of behavioral regularities and convergent expectations on the part of their members. While such institutional arrangements are difficult to alter in a planned or guided fashion, they change continuously in response to both their own inner dynamics and a variety of political, economic, and social factors in their environments. This fact suggests the importance of posing two sets of questions regarding regime dynamics. How and why do regimes arise from the interactions of individual actors over time? The argument of this chapter is that it is helpful to differentiate three developmental sequences in connection with regime formation at the international level. The resultant institutional arrangements can be labeled spontaneous regimes, negotiated regimes, and imposed regimes. Additionally, we want to know how and why regimes change once they become fully established. Here it seems illuminating to distinguish between endogenous forces involving internal contradictions of one sort or another and exogenous forces involving shifts in underlying power structures, prevailing technology, human values or tastes, and so forth. If we adopt this perspective, the next task in the study of regime dynamics is to seek a more sophisticated understanding of the factors governing the incidence of these developmental sequences and patterns of change.

Part Two

INTERNATIONAL REGIMES
IN PRACTICE

Prologue

We have long been accustomed to studying the constraints that biological and physical systems place on the course of human affairs. Among those interested in international relations, studies of such constraints have typically taken the form of efforts to link the power of states to measures of their natural-resource endowments and, somewhat more grandiosely, of attempts on the part of geopoliticians to explore the strategic consequences of geographical or spatial configurations. Recently, however, we have begun to realize that human activities, in the aggregate, can produce dramatic impacts on natural systems and, in the process, affect the habitability of the planet for future generations. This became apparent initially in connection with the observation that human harvesting has played a major role in the depletion of fish stocks and other renewable resources, such as bison, whales, and seals. Increasingly dramatic examples have now come into focus. Even if the more extreme versions of the nuclear winter scenario are wide of the mark, it is undeniable that nuclear warfare could drastically reduce the habitability of the planet for human beings. And the environmental consequences of less apocalyptic occurrences, such as the depletion of stratospheric ozone caused by chlorofluorocarbon emissions or the global warming trend associated with the buildup of carbon dioxide, may prove almost equally disruptive in the long run. With remarkable speed, therefore, issues pertaining to natural resources and the environment have appeared on the agenda alongside the more traditional issues of defense or security and economics as major preoccupations of students of international relations. If anything, these newly emergent environmental issues are likely to move closer to center stage during the foreseeable future.

Given the role of institutional arrangements as important determinants of collective outcomes in international society, the fact that we are now witnessing a growing interest in international regimes for natural resources and the environment seems entirely appropriate. The analysis of such regimes should allow us to understand when and why human behavior will prove disruptive to the biological and physical systems that are important in sustaining human life on the planet. Additionally, such an analysis must be pursued vigorously if we are to have any hope of safeguarding critical natural systems against disruptive impacts from human activities. A better understanding of the consequences of institutional arrangements for key natural systems is a necessary (though not sufficient) condition for efforts to devise or revise international regimes in such a way as to control the adverse effects of human activities or to redirect these activities into less destructive channels.

The chapters of Part 2 offer some illustrative examples of the use of regime analysis to deal with international issues relating to natural resources and the environment. Chapter 5, which offers a form of comparative statics, addresses relatively conventional concerns relating to renewable resources (fish) and nonrenewable resources (hardrock minerals). Turning to the dynamics of regime formation, chapter 6 focuses on the nuclear-accident problem and the associated issue of handling transboundary movements of radioactive fallout. But similar transboundary problems arise in connection with threats to biological diversity, ozone depletion, and the global warming trend. Chapter 7 proceeds from the growing recognition that large natural systems commonly provide environmental services in contrast to the commodities we generally envision in using the concept of natural resources. While the chapter emphasizes the services of marine systems in providing sea lanes for commerce, other environmental services, such as the absorption of wastes or residuals or the transmission of radio and television signals, now require increased attention in international society.

CHAPTER FIVE

Comparative Statics: Regimes for the Marine Fisheries and Deep-Seabed Mining

Some natural resources lie outside the bounds of national jurisdictions or cut across existing jurisdictional boundaries in such a way that effective management by individual states is not feasible. The principal concentrations of manganese nodules in the Pacific Ocean, for example, are in areas that lie beyond even the most expansive jurisdictional claims of coastal states. Many stocks of fish and other oceanic resources, such as marine mammals or oceanic birds, range over extensive areas without regard for the jurisdictional boundaries of sovereign states. By its nature, maritime commerce relies on sea-lanes passing through two or more sovereign jurisdictions. Similarly, space resources, such as the electromagnetic spectrum employed in radio and television broadcasting, are not subject to effective control by individual states.

Sometimes jurisdictional boundaries can be restructured to bring natural resources effectively under the control of individual states. In such cases, it is pertinent to examine the relative merits of establishing international regimes featuring well-defined zones within which individual states exercise management authority in preference to open-to-entry arrangements or institutions involving supranational rules and administrative agencies. In other cases, there is little prospect of bringing the relevant resources wholly under the jurisdiction of individual states (for example, deep-seabed minerals, oceanic birds). Under these conditions, the real choice with regard to management is between open-to-entry utilization (that is, unrestricted common-property regimes) and some sort of supranational arrangements (that is, systems of rules and administrative agencies that transcend the jurisdiction of individual states).[1]

---

1. Taken together, regimes featuring national zones, open-to-entry common property, and supranational arrangements compose the domain of international regimes, as that concept is used throughout this book.

Without doubt, ambiguous cases falling between these extremes are common under real-world conditions. Attempts on the part of individual states to extend their jurisdiction may achieve only partial success. The current American effort to gain full control over commercially harvested salmon of North American origin in the North Pacific, for instance, will probably fall into this category. Similarly, the introduction of supranational arrangements often proves ineffectual in practice. Such a fate has befallen a number of the regional regimes set up to manage marine fisheries. And some thoughtful observers have expressed concerns along these lines regarding the proposed International Seabed Authority for deep-seabed mining.

Even allowing for such intermediate cases, however, these observations suggest the importance of asking several distinct questions about institutional arrangements or regimes pertaining to natural resources at the international level. Are supranational arrangements desirable in connection with specific natural resources? The critical issue here concerns the merits of supranational arrangements in comparison with other possible regimes for the same resource. In those cases where it does seem desirable to establish supranational regimes, what specific institutional arrangements seem most promising? The issue here is essentially a matter of devising social institutions to accommodate the ecological, economic, and political features of the resource management problem at hand.

This chapter offers an exercise in comparative statics. It examines the relative merits of alternative institutional arrangements, without reference to the politics of regime formation.[2] For the sake of concreteness, the following discussion draws heavily on the cases of marine fisheries and deep-seabed mining. But the underlying issues are generic; they arise throughout the international relations of resource management and beyond. The central thesis of the chapter is that the specific provisions of international regimes typically have far-reaching consequences both for the allocation of factors of production and for the distribution of wealth.

Are Supranational Institutions Desirable?

It is by no means self-evident that supranational arrangements are desirable in the management of any given natural resource in interna-

2. For a detailed account of criteria of evaluation in terms of which to judge efforts to devise institutional arrangements see Oran R. Young, *Resource Regimes: Natural Resources and Social Institutions* (Berkeley: University of California Press, 1982), chap. 5.

tional society. Even when resources are distributed in such a way that they cut across the existing bounds of national jurisdictions, there are apt to be management options that do not require the introduction of detailed rules and administrative agencies transcending the jurisdiction of individual states. The task of this section, then, is to examine the conditions under which the creation of supranational arrangements is a preferred strategy in the international relations of resource management. To lend substance to the discussion, I focus on the case of marine fisheries.[3]

There is general agreement that unrestricted common-property or open-to-entry arrangements will yield highly unfortunate results in the marine fisheries under contemporary conditions.[4] The major problems include overfishing and the resultant depletion of fish stocks, the dissipation of economic returns or rents, overcapitalization in the form of expensive gear that lies idle much of the time, and low wages and underemployment in the pertinent labor force. A rising demand for fish products worldwide has ameliorated some of the consequences of this situation in purely economic terms. But it has done nothing to alter the fact that unrestricted common-property arrangements are fundamentally inadequate for the management of marine fisheries.[5] To be precise, this proposition applies with full force only to cases where fish stocks are subjected to heavy usage. But this restriction is not of great importance analytically since heavy usage has become the rule rather than the exception in the marine fisheries, and there is every reason to expect that worldwide demand for fish products will continue to rise. In consequence, many commentators now argue that almost any alternative is preferable to unrestricted common-property regimes for the marine fisheries.

What are the alternatives to unrestricted common property in this realm? While many variants are possible, the principal options fall into three groups. The first involves an extension of coastal-state jurisdiction coupled with the establishment of fishery-conservation zones on the part

3. The stakes in the marine fisheries are high. In 1975, for example, worldwide landings of fish from the marine fisheries amounted to 69.7 million metric tons (live weight). The American catch in 1975 was 2.8 million metric tons, with a total value of $971 million or an average value of $347 per ton. See Frederick W. Bell, *Food from the Sea: The Economics and Politics of Ocean Fisheries* (Boulder: Westview Press, 1978), 362.

4. For a seminal account see H. Scott Gordon, "The Economic Theory of a Common Property Resource: The Fishery," *Journal of Political Economy* 62 (1954), 124–142.

5. As James Crutchfield puts it, "The basic theory of a high sea fishery, whether exploited by a single nation or by more than one nation, suggests a bleak economic existence, to say the least" ("The Marine Fisheries: A Problem in International Cooperation," *American Economic Review* 54 [1964], 212).

of these states. A second option rests on the idea of creating a global system, including an administrative apparatus with sufficient power and authority to manage the marine fisheries of the world on an integrated basis. An intermediate option features a collection of regional arrangements demarcated in terms of distinct ecosystems. Selecting an alternative from the first group implies that supranational arrangements are not needed. The choice of a regime from either of the other groups, by contrast, suggests a substantial role for supranational arrangements.

Coastal-State Jurisdiction

In fact, we have recently experienced a marked trend toward extended coastal-state jurisdiction in the marine fisheries. Yet this trend does not mean that this institutional option will produce attractive solutions for the problems of unrestricted common property. Though it is surely too soon to formulate definitive conclusions concerning this sort of regime, an examination of the arrangements that have been put in place under the provisions of the American Fishery Conservation and Management Act of 1976 (FCMA) makes it possible to identify and analyze certain reservations about regimes of this type.[6]

The FCMA establishes a fishery-conservation zone (Sec. 101) within which the United States asserts exclusive management authority (Sec. 102). This zone extends "200 nautical miles from the baseline from which the territorial sea is measured," except in the case of anadromous species for which management authority is claimed "throughout the migratory range of such species." Though the impact of this effort to expand coastal-state jurisdiction is substantial, it is less drastic than it may at first appear. The FCMA does not attempt to extend exclusive management authority to cover highly migratory or pelagic species (for example, tuna). It merely reaffirms the already emerging situation with respect to sedentary species like crabs or oysters. And it is doubtful whether the claims of the FCMA regarding extended jurisdiction over anadromous species can be implemented fully and effectively in such critical cases as the salmon of the North Pacific. The major change heralded by the FCMA therefore relates to the large stocks of demersals or groundfish located within the 200-mile limit. Here it is indeed reasonable to expect that the Act will lead to fundamental changes in the preexisting situation.

The FCMA sets up an elaborate organizational apparatus to carry out

6. Fishery Conservation and Management Act of 1976, PL 94–265, 90 Stat. 331 codified at 16 USCA 1801–1882.

the responsibilities arising from the assertion of exclusive management authority in the marine fisheries off the coasts of the United States. The distinctive feature of this apparatus is a system of eight regional fishery-management councils (Sec. 302). These councils are responsible for the formulation of detailed fishery-management plans (Sec. 303), and they are well positioned to play a major role in the implementation of these plans with respect to specific fisheries. Yet the Act also allocates a major role in fisheries management to the Secretary of Commerce as represented by the National Marine Fisheries Service (NMFS). The NMFS is expected in many areas to operate either in conjunction with the regional councils or on some concomitant basis (Sec. 304). In fact, it is far from clear where effective power lies in this system.[7] It will undoubtedly require considerable experience with this regime to determine the nature of the balance of power between the regional councils and the NMFS.

Turn now to the problems posed by extended coastal-state jurisdiction. It is a little simplistic but not, I think, inaccurate to say that the basic problem of conservation in the marine fisheries is to prevent severe stock depletions arising from overfishing. In open-to-entry fisheries, this problem stems from the intrinsic characteristics of unrestricted common property. Where conscious regulation is present, by contrast, the central problem is to determine total allowable catches (TACs) on an annual basis and to administer them effectively. The FCMA deals with this issue by establishing the criterion of *optimum yield* (Sec. 3 [18]). Unlike the concepts of maximum sustained yield (MSY) and maximum economic yield (MEY), however, the idea of optimum yield has no real analytic content. As Larkin puts it, "Optimum yield is whatever you wish to call it."[8] It is true that the FCMA appears to designate maximum sustained yield as a sort of baseline in the computation of TACs.[9] Nonetheless, this is of little practical significance both because the Act envisions numerous departures from this baseline and because the criterion of MSY itself is notoriously difficult to operationalize under real-world conditions (especially in cases involving multiple-species fisheries). It

7. The FCMA also allocates significant roles to the secretary of state under Secs. 204 and 205 as well as to the U.S. Coast Guard under Sec. 311.

8. Peter A. Larkin, "An Epitaph for the Concept of Maximum Sustained Yield," *Tranactions of the American Fisheries Society* 106 (January 1977), 8.

9. Sec. 3 (18) of the FCMA states specifically that "[t]he term 'optimum,' with respect to the yield from a fishery, means the amount of fish—(A) which will provide the greatest overall benefit to the Nation, with particular reference to food production and recreational opportunities; and (B) which is prescribed as such on the basis of the maximum sustainable yield from such fishery, as modified by any relevant economic, social, or ecological factor."

follows that the concept of optimum yield "is potentially subject to abuse, and [will] almost certainly be used as a way of justifying a political course of action."[10] To say the least, this is not reassuring. The decision-making processes set up under the FCMA are weighted heavily toward the interests of the domestic fishing industry. And this is an industry that can hardly be said to have an impressive record of concern for its own long-term interest in the viability of fish stocks, much less for the preservation of healthy stocks of fish as a goal in its own right.

What of the implications of the FCMA regime for allocative efficiency?[11] To begin with, the Act is vague on the extent to which efficiency is to be treated as an important goal in the management of marine fisheries. Optimum yield is not fundamentally an economic criterion; economic considerations constitute only one of a list of factors to be included in computing optimum yield. When it does turn more directly to economic guidelines, the Act sets forth a rather ambiguous standard to the effect that "[c]onservation and management measures shall, where practicable, promote efficiency in the utilization of fishery resources; except that no such measure shall have economic allocation as its sole purpose" (Sec. 301 [a] [5]).

It is probable that the implementation of the FCMA will produce at least a temporary reduction in the annual harvests of some stocks of groundfish (for example, Alaskan pollock) due to limitations on foreign fishing.[12] This reduction will yield de facto improvements in economic efficiency in the short run.[13] But the FCMA does not take a strong stand on the introduction of effective entry restrictions, generally regarded as the key to the achievement of efficiency in the marine fisheries.[14] Entry restrictions are relegated to the category of "discretionary provisions"

10. Larkin, "Maximum Sustained Yield," 8.

11. In principle, the concept of efficiency is sufficiently general to encompass all benefits and costs, however conceptualized. But most analyses confine the domain of efficiency, in practice, to those benefits and costs arising from private-good transactions or from other situations where "prices" are easily assignable. The resultant computations typically ignore many nonmarket phenomena such as social costs (and benefits), public goods, and various political conditions. For a helpful discussion of the concept of economic or allocative efficiency see Robert Dorfman and Nancy Dorfman, eds., *Economics of the Environment*, 2d ed. (New York: Norton, 1977), 7–25.

12. But note that there is no intention of terminating foreign fishing abruptly. See Robert J. Samuelson, "New Fishing Limits May Prove Big Haul," *Washington Post*, 29 November 1977, D7–8, and "U.S. Trims Foreign Catch," *Washington Post*, 29 November 1977, A3.

13. This result is a consequence of the fact that the single most important source of inefficiency in the marine fisheries has been overinvestment arising from the common-property character of the resource.

14. Francis T. Christy and Anthony Scott, *The Common Wealth in Ocean Fisheries* (Baltimore: Johns Hopkins University Press, 1965).

under Sec. 303 (b). Numerous factors are to be taken into consideration if and when entry restrictions are established, including a catchall factor called "any other relevant considerations." Moreover, Secs. 303 and 304 are written in such a way that the subject of entry restrictions is likely to become a bone of contention between the regional councils and the NMFS.[15] This fact suggests that political considerations, in contrast to the criterion of allocative efficiency, will dominate any effort to devise entry restrictions. And there is no reason to expect that the play of politics will produce results resembling those that would be required to promote efficient outcomes.

Additionally, the FCMA virtually guarantees the introduction of a new source of allocative inefficiency in the marine fisheries of the American zone. It calls for efforts to expand the role of high-cost American operators and to reduce the role of more efficient foreign operators in these fisheries (for example, Sec. 2 [7]).[16] In short, the Act establishes a form of protectionism for the marine fisheries of the fishery-conservation zone. While the actual impact of this feature of the regime will depend upon the way in which the Act is administered in specific regions, there is no escaping the fact that it involves a movement away from rather than toward allocative efficiency.[17] I do not find the arguments supporting protectionism (for example, guaranteeing employment, helping an industry in temporary trouble, promoting national security) compelling in this case. It is true that this feature of the FCMA is compatible with the concurrent rise of economic protectionism in other areas and may therefore seem acceptable to some. At a minimum, however, those who take this position should accept the political implications of their views explicitly rather than cloak them under the banner of conservation.

I do not find the distributive implications of the FCMA regime any more reassuring, though this is admittedly an area where value judgments are unavoidable. The initial effect of the FCMA will certainly be to redistribute some of the proceeds from the marine fisheries to coastal-

15. The key point is that both the regional councils and NMFS are authorized to work on fishery management plans. Though there is reason to expect the interests of the two groups to conflict from time to time, the FCMA does not indicate clearly how such conflicts are to be resolved. See also U.S. Congress, Office of Technology Assessment, *Establishing a 200-mile Fishery Zone* (Washington, D.C.: GPO, 1977).

16. Sec. 2(7) of the FCMA, for example, states that "[a] national program for the development of fisheries which are underutilized or not utilized by United States fishermen, including bottom fish off Alaska, is necessary to ensure that our citizens benefit from the employment, food supply, and revenue which could be generated thereby."

17. For evidence regarding the initial approach to administering the FCMA see Samuelson, "New Fishing Limits."

state (though not necessarily local) fishermen.[18] No doubt, the federal government could intervene to alter this initial effect through taxation, price setting, or transfer payments, but the probability of any coherent initiatives along these lines is low. This raises several more specific distributive issues that are troublesome. Where foreign operators are engaged in well-established fisheries in the fishery-conservation zone (for example, Japanese operators in the Bering Sea), the Act promotes what amounts to a form of expropriation without compensation.[19] In this instance, it is use rights rather than fixed assets that are subject to expropriation, but this hardly makes the shift more palatable.[20] Similarly, there are problems of equity in the allocation of permits for foreign fishing as well as permits for domestic fishing in the event that meaningful entry restrictions are introduced.[21] The Act leaves the allocation of permits for foreign fishing to an open-ended administrative process (Sec. 201 [e]), and it is virtually silent on processes for allocating allowable catches among domestic fishermen. It is reasonable to assume that outcomes in this realm will be influenced heavily by the actions of the regional councils. As I have already suggested, the regional councils are politicized entities, and there is no basis for assuming that their decisions will conform to reasonable standards of equity or justice. Should the regime introduced under the FCMA succeed to the point where it is possible to reap significant economic returns from fish stocks, moreover, questions will arise regarding the distribution of these returns. The FCMA is silent on the issue. It does not set up a procedure (for example, a leasing system for the use of fishing grounds) under which some of the proceeds derived from the exploitation of fish stocks could find their way into public hands. This suggests that the state is not viewed as a proper beneficiary from the harvest of fish (in contrast to the

18. Ironically, the FCMA provides no real guarantees for truly local (and traditional) fishermen. The offshore fisheries of Alaska, for example, might well come to be dominated under the regime by Seattle-based firms (some of which are partially owned by Japanese investors).

19. A possible response would be for foreign interests to invest directly in American fishing operations or processing plants. Some direct investment of this type has already occurred. But it is politically sensitive, and the American government is likely to come under heavy pressure to restrict investment of this type. For some illustrative data, see Per O. Heggelund, "Japanese Investment in Alaska's Fishing Industry," *Alaska Seas and Coasts* 5 (October 1977), 1–2, 8–9.

20. For more general discussions of property rights and interests see Harold Demsetz, "The Exchange and Enforcement of Property Rights," *Journal of Law and Economics* 7 (1964), 1–26, and J. H. Dales, *Pollution, Property, and Prices* (Toronto: University of Toronto Press, 1968).

21. Given limited entry, permits conveying rights to fish become factors of production in their own right. There are numerous methods of allocating such rights, and these methods can be expected to produce widely differing consequences.

exploitation of oil) in marine areas.[22] Given the current unprofitable condition of the fishing industry, this problem may seem academic. But the distributive implications of such a posture are hard to justify in any persuasive way.

Equally troublesome is the fact that the FCMA treats the management of marine fisheries in a compartmentalized fashion. That is, it fails to recognize interdependencies among marine resources (for example, fish stocks, marine mammals, hydrocarbons, sea lanes, marine ecosystems), and it makes no effort to fit the management of the marine fisheries into a larger set of ocean policies. Yet it is apparent that we are moving toward heavier usage of many marine resources and that conflicts among marine activities are already assuming a prominent place in the arena of public policy. To take a specific example, how are we to resolve conflicts between the interests of fishing operators and the interests of those desiring to proceed as rapidly as possible with the extraction of oil from the outer continental shelves?[23] I have no general solution to suggest for these problems; they raise fundamental political questions that cannot be resolved through the use of technical procedures like benefit/cost analysis.[24] Nonetheless, it seems clear that any regime that focuses on individual resources in a compartmentalized fashion leaves much to be desired under contemporary conditions.

Finally, it is worth emphasizing that the implications of the FCMA are unfortunate from the perspective of promoting international cooperation in an increasingly interdependent world. Viewed in international terms, the Act appears to license a unilateral redistribution of rights without compensation.[25] It is simply not true that the relevant stocks of fish were previously unowned or not subject to legitimate property interests (that is, res nullius) so that the promulgation of exclusive management authority in the marine fisheries amounted to the asser-

22. The question is whether the state as owner or manager of a valuable factor of production is entitled to receive normal economic returns from the use of this factor. For the analogous case of outer continental-shelf oil and gas see J. W. Devanney III, *The OCS Petroleum Pie*, MIT Sea Grant Program, Report No. MITSG 75–10 (Cambridge, 1975).

23. To take a concrete case, consider the controversy that arose in the 1970s over the oil and gas lease sale in Kachemak Bay in the Lower Cook Inlet of Alaska. Under severe pressure from fishing interests, the State of Alaska actually bought back the oil and gas leases it had sold in this area. See also Nancy Munro, "OCS Development—What It Means," *Alaska Seas and Coasts* 3 (April 1975), 1–4.

24. In effect, the problem turns on what should be conceptualized as benefits and costs to begin with, quite apart from the procedural difficulties involved in assigning numerical values to specific benefits and costs.

25. This proposition is true despite the fact that the United Nations Conference on the Law of the Sea has been moving toward sanctioning expanded coastal-state jurisdiction over the marine fisheries. See also Robert J. Samuelson, "Law of the Sea Treaty—Talk, Talk, and More Talk," *National Journal* 35 (27 August 1977), 1337–1343.

tion of rights where none existed before.[26] In fact, this action entailed the extinguishment of existing rights together with the introduction of new rights on a unilateral basis. What is more, it is hard to escape the conclusion that this transition has been effected in a harsh and discriminatory fashion. Let me illustrate this point. The punishments for criminal offenses set forth in the FCMA (Sec. 309) discriminate sharply between American fishermen and foreign fishermen. The Act enunciates a discriminatory position toward historic or traditional fishing rights, appearing to ignore them where the activities of foreign fishing operators are concerned (Sec. 201 [d]) while insisting that they be recognized in the case of American operators (Sec. 202 [e]). The Act articulates what turns out to be a self-serving position on the regulation of highly migratory species. In fact, the phrase "highly migratory species" is construed essentially to mean tuna (Sec. 3 [14]), and the Act goes to some lengths to deter foreign interference with this important American fishery (for example, Sec. 205 [a]). Furthermore, the FCMA lays out an extraordinarily complex system of administrative procedures for the issuance of permits for foreign fishing (Sec. 204 [b]), which cannot fail to curtail the operations of foreign fishermen beyond what is contemplated under the terms of Sec. 201 (d). The contrast between this bureaucratic maze and the relative absence of such complications for domestic fishermen (Sec. 303 [b] [1]) is striking.

The point of mentioning these details is straightforward. In a highly interdependent world, regimes that not only impose redistributions of rights in a unilateral fashion but also do so in a discriminatory fashion have dangerous implications. Other states with the capacity to do so (for example, Japan) are likely to consider retaliating in various ways. Retaliatory actions need not, of course, be limited to the realm of marine fisheries; they may well involve other economic or even political concerns. Additionally, American positions on other issues relating to marine policy (for example, freedom of navigation, flag-state predominance in dealing with marine pollution, freedom of scientific research) are likely to be received with growing coolness on the part of other states. More generally, discriminatory unilateralism raises the specter of a new round of protectionist moves that could lead to economic warfare. Though it is not clear that coastal-state regimes for the marine fisheries need exhibit this drawback, such concerns are surely troubling in con-

26. For a review of the more general literature on property rights see Eirik Furubotn and Svetozar Pejovich, "Property and Economic Theory: A Survey of Recent Literature," *Journal of Economic Literature* 10 (1972), 1137–1162.

nection with the regime laid out in the Fishery Conservation and Management Act of 1976.

A Global Regime

It follows from the preceding discussion that both unrestricted common-property arrangements and zonal arrangements based on coastal-state jurisdiction have serious limitations as regimes for the marine fisheries under contemporary conditions. This conclusion suggests that the options for supranational arrangements are at least worthy of serious examination at this stage. Turn, then, to the idea of a global arrangement, including some supranational administrative apparatus capable of managing the world's fisheries on a unified basis. This might amount to something like the Department of Fisheries of the United Nations Food and Agriculture Organization, though endowed with more effective power and authority. But numerous variations on this institutional theme are possible.[27] Even so, it seems feasible to arrive at some general propositions about any regime of this sort without examining the variations in detail.

Without doubt, this option runs straight into the objection that it would be politically unacceptable to many influential members of international society. Some states would refuse to agree, even on paper, to a global regime with sufficient power and authority to manage the marine fisheries of the world. Additionally, there are good reasons to suspect that the administrative apparatus of such a regime would prove ineffectual in practice, even if states were to subscribe to such an arrangement in principle. This conclusion is not, however, particularly distressing in the case of the marine fisheries. It is comparatively easy to show that the creation of a global regime would not constitute the preferred option for managing the marine fisheries, even it it were feasible.

Global arrangements are not required to achieve conservation or allocative efficiency in most of the world's fisheries. In fact, the typical fishery involves a regional, as opposed to global, commons.[28] While the relevant ecosystems are often difficult to subsume under the jurisdiction of individual states, they do not begin to approach the global level in

27. Edward Miles, *Organizational Arrangements to Facilitate Global Management of Fisheries*, Resources for the Future Program of International Studies of Fisheries Arrangements, Paper No. 4 (Washington, D.C.), 1974.
28. For a selection of essays on various types of commons see Garrett Hardin and John Baden, eds., *Managing the Commons* (San Francisco: W. F. Freeman, 1977).

most cases. This is true even if we take a highly sensitive view of interdependencies among intersecting fisheries. There are certainly exceptions to this proposition. The obvious example involves whales, where management is clearly a worldwide concern. A somewhat less impressive case can be made for treating tuna as an exception.[29] Efforts to manage these fisheries on a global basis may well be justified, but the general rule is that neither conservation nor allocative efficiency requires extending the scope of management arrangements beyond the boundaries of the relevant ecosystems.

Nor am I persuaded that considerations of equity or justice require a global regime for the marine fisheries. This is partly a consequence of the fact that such a regime would be ineffectual, failing to produce clear-cut outcomes in many instances. As a result, it would often become embroiled in activities tending to reduce the size of the pie to be distributed rather than concentrate on distributing obtainable proceeds from the marine fisheries in an equitable fashion. Until recently, at least, the global regime for whaling exemplified this failing.[30] In part, however, the problem stems from the fact that there is no guarantee that a global administrative apparatus would produce equitable outcomes in the allocation of total allowable catches or the distribution of economic returns from the marine fisheries. Decisions on such matters would inevitably involve a bargaining process played out in a highly politicized environment. In many ways, they would resemble the negotiations during the 1970s and 1980s in the context of the Third United Nations Conference on the Law of the Sea. At this stage, it is hard to muster a feeling of confidence that such processes would yield results that conform to even the most modest standards of equity or justice.

Should any doubts remain at this juncture, the problem of transaction costs also looms in connection with the option of a global regime for the marine fisheries. Such costs (including decision costs, administrative costs, and compliance costs) must be taken into account in assessing the merits of any regime for the marine fisheries other than unrestricted common-property arrangements. But there is general agreement that transaction costs tend to increase rapidly as a function of the scale of institutional arrangements (measured in terms of geographical domain,

29. Saul B. Saila and Virgil J. Norton, *Tuna: Status, Trends, and Alternative Management Arrangements,* Resources for the Future Program of International Studies of Fishery Arrangements, Paper No. 6 (Washington, D.C.), 1974.

30. On the earlier history of the international whaling regime see George L. Small, *The Blue Whale* (New York: Columbia University Press, 1971).

functional scope, and size of membership).[31] It seems likely, therefore, that a truly global regime for the marine fisheries would be prohibitively expensive to operate. The transaction costs alone might well reach a level where they would preclude the promulgation of any well-defined regime of this sort. This is one way to view the problems that plagued the law-of-the-sea negotiations, and I can see no reason to conclude that a global regime for the marine fisheries would fare better. It seems hard to escape the conclusion that supranational institutions in the form of a comprehensive global regime are not a preferred option for the management of the world's marine fisheries.

Regional Arrangements

There remains for consideration the other group of options involving supranational institutions. The essential idea here would be to establish a collection of regional regimes, each of which would assume responsibility for the management of the marine fisheries of a geographically demarcated area. Each arrangement would take the form of what Crutchfield calls a "supranational fishing entity with exclusive rights over a geographic area that encompasses an appropriate ecological unit."[32] While it is certainly true that many variations on this theme are also possible, let me endeavor to articulate some general propositions about regimes of this sort for the marine fisheries.

Prior experience with regional regimes for marine fisheries surely leaves much to be desired. Yet it would be a mistake to draw unduly pessimistic conclusions from this observation. Some regional arrangements have undoubtedly produced constructive results. Examples such as the North Pacific halibut regime, the Fraser River salmon regime, and the fur-seal regime come to mind in this connection.[33] Many existing regional arrangements, moreover, have operated under distinctly unfavorable conditions. Either the members have failed to provide the administrative apparatus with adequate authority to begin with, or one or more of the members have attempted to exploit the regime to impose some preferred outcome on the others. Experience with regimes such as the one laid out in the International North Pacific Fisheries Convention

31. E. J. Mishan, "The Postwar Literature on Externalities: An Interpretive Essay," *Journal of Economic Literature* 9 (1971), 21–24.
32. Crutchfield, "Marine Fisheries," 214–215.
33. See also Christy and Scott, *Common Wealth in Ocean Fisheries.*

reflects both these problems.[34] Consequently, simple extrapolations from past experience are hazardous in this realm, though they do serve to sensitize us to the political problems inherent in efforts to establish effective supranational institutions for the management of any natural resource.

In fact, regional arrangements have substantial advantages as regimes for the marine fisheries. They surely constitute, in principle, the preferred option from the perspectives of conservation and allocative efficiency. In contrast to a global regime as well as coastal-state jurisdiction, regional arrangements can be adjusted to make their boundaries congruent with the ecosystems and economic systems involved in complexes of marine fisheries.[35] Thus, the geographical domain of any given regional regime could be extended to encompass the range of the relevant fish stocks. Similarly, functional scope could be tailored to take account of interdependent fisheries (for example, the groundfisheries of the Bering Sea or the Northwest Atlantic). And it would be possible to include as members all states with a serious interest in the relevant fisheries without (in most cases) running into the problems of collective action in large groups.[36] Under the circumstances, it ought to be feasible to endow individual regional regimes with organizations possessing adequate authority to promote conservation and allocative efficiency in their respective areas. No doubt, this would require the introduction of systems of entry restrictions and reasonable procedures for dealing with new entrants, but there are no insuperable technical problems in this realm. Whether most states would find regimes of this type politically acceptable is another matter. Past experience certainly does not license any easy optimism in this connection. But it is worth noting that we are now witnessing the emergence of a serious interest in trying regional arrangements for some fisheries around the world. Thirteen nations joined in negotiating the regional regime set forth in the 1980 Convention on the Conservation of Antarctic Marine Living Resources, for example, and twelve South Pacific states have taken steps to create a regional arrangement to manage their combined fisheries zones.[37]

34. Hiroshi Kasahara and William Burke, *North Pacific Fisheries Management*, Resources for the Future Program of International Studies of Fishery Arrangements, Paper No. 2 (Washington, D.C.), 1973.

35. Crutchfield, "Marine Fisheries," 215.

36. The seminal modern work on collective-action problems is Mancur Olson, Jr., *The Logic of Collective Action* (Cambridge: Harvard University Press, 1965).

37. Bernard D. Nossiter, "Antarctic Nations Move to Control Fishing in Region," *Washington Post,* 30 September 1977, A22, and "South Pacific States Set Up 200-Mile Limits," *Christian Science Monitor,* 25 October 1977, 21.

There is also a case for regional regimes in terms of the criterion of equity or justice. Such matters as the allocation of shares of allowable catches and the handling of economic returns would certainly give rise to hard bargaining within each regional authority, and I do not mean to assert that the outcomes would conform to some ideal standard of fairness in every case. Yet no individual member could expect to occupy the driver's seat regarding distributive issues, as is inevitably the case with zonal regimes featuring coastal-state jurisdiction. In most cases, moreover, regional regimes (unlike a global arrangement) would be manageable enough to permit a reasonable expectation that bargaining would yield outcomes falling on something like a welfare frontier rather than produce a drift toward inconclusive or Pareto inferior results of the type that commonly occur when the number of participants is too large. It is true that regional arrangements might tend to slight the just claims of new entrants from outside the original group as well as of consumers of fish products worldwide. Though efforts could be made to combat these tendencies (for example, through the establishment of transferable fishing rights), it is hard to imagine eradicating them altogether. In my judgment, however, these distributive drawbacks are mild compared with the distributive problems of the other types of regimes.

Consider, finally, the question of transaction costs associated with regional arrangements. Without doubt, these costs would be significant, and they would tend to increase at the margin as any regional regime expanded in group size, geographical domain, or functional scope. It is surely fair to assume that declining marginal returns would set in at some point from such expansions. Without attempting to operationalize these concepts here, this scenario does suggest some simple conclusions.[38] There will be an optimal size for any regional fisheries regime. This size (defined in terms of some combination of group size, geographical domain, and functional scope) will occur where the marginal gains from expansion are just matched by the marginal costs (defined in terms of transaction costs). The exact point at which optimal size is reached will be determined by the shapes of the marginal-gains and marginal-costs curves. In Figure 2 optimality occurs at the point labeled X_i. Manipulations of the shapes of these curves can lead to great variations in assessments of the appropriate size for any given regional regime (including the conclusion that a regional regime should not be set up at all). I think it is fair to conclude, however, that such assessments

---

38. For a somewhat similar argument see Todd Sandler and Jon Cauley, "The Design of Supranational Structures," *International Studies Quarterly* 21 (1977), 251–276.

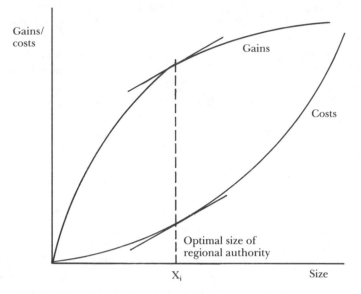

Figure 2. Optimal size of regional resource-management authority

will generally support the proposition that regional regimes make good sense for the marine fisheries under contemporary conditions.

To return to the basic question of this section, are supranational arrangements desirable in efforts to manage natural resources at the international level? The preceding argument does not license any simple answer to this question. It is illusory to expect any regime to yield results in practice that conform perfectly to some ideal standards of performance. Nevertheless, the argument of this section is instructive with regard to our basic question. There is general agreement that unrestricted common-property arrangements are inadequate for most marine fisheries. Zonal arrangements featuring coastal-state jurisdiction also leave much to be desired not only on grounds of justice or equity but also on grounds of conservation and allocative efficiency. This suggests that regimes involving supranational institutions are worthy of serious consideration in connection with the marine fisheries. While a global regime would be fraught with problems in this realm, regional arrangements for the marine fisheries exhibit a number of attractive features. Accordingly, it seems desirable to proceed now to an analysis of major issues that arise when we consider the relative merits of alternative supranational arrangements for the management of natural resources.

TYPES OF SUPRANATIONAL INSTITUTIONS

The conclusion that supranational arrangements are desirable in managing any given natural resource opens up a new range of questions about international regimes. An array of supranational alternatives are available with regard to most natural resources, a proposition that takes on significance in light of the fact that the selection of a particular set of arrangements will ordinarily have far-reaching consequences in terms of social as well as private benefits and costs. The objective of this section, therefore, is to explore the nature and consequences of detailed institutional options, with special reference to the formation of supranational institutions for the management of natural resources. This time, I focus on deep-seabed mining. Doing so will broaden the scope of our enquiry into international-resource regimes. At the same time, the principal implications of the analysis should apply to the marine fisheries and other natural resources as well.

Some have argued that an unrestricted common-property regime for deep-seabed mining would be perfectly acceptable, and perhaps even desirable, from the perspective of a country like the United States.[39] The case for this view involves the following line of reasoning.[40] There are fundamental differences between manganese nodules on the one hand and stocks of fish or pools of oil on the other. It is not apparent that the absence of exclusive rights to manganese nodules would reduce the allocative efficiency of deep-seabed mining by encouraging either the underinvestment that arises from free-rider problems or the overinvestment that arises from common-pool problems. As Eckert puts it, economic analysis suggests that "a policy favoring regulation cannot be supported on grounds of economics until more is known about the

39. It is not easy to calculate the stakes involved in deep-seabed mining. Estimates of the magnitude of the nodules themselves are highly speculative and vary widely; the range 1.5 to 3 trillion tons is mentioned frequently. Estimates of the value of the minerals recoverable from the nodules are even harder to come by since all such estimates are nothing but projections of future costs and revenues, and numerous factors (both economic and political) may operate to alter current estimates drastically before hard minerals from the deep seabed come on the market. It is interesting to note, however, that "the mining firms estimate cumulative expenditures of at least $109 million to develop deep seabed mining technology and cumulative investments of from $1.5 to $2.4 billion to achieve commercial operations" (U.S. Congress, Office of the Comptroller General, *Deep Ocean Mining Environmental Study: Information and Issues* [Washington, D.C.: GPO, 1977], 5).

40. This argument is expressed with clarity and force in Ross D. Eckert, "Exploration of Deep Ocean Minerals: Regulatory Mechanisms and United States Policy," *Journal of Law and Economics* 17 (1974), 143–177. See also the more extended account in Ross D. Eckert, *The Enclosure of Ocean Resources* (Stanford: Hoover Institution Press, 1979).

industry's operating conditions, costs, and incentives."[41] To the extent that this argument holds, there is no reason to assume that the investment climate for deep-seabed mining would be severely affected by the absence of exclusive rights in nodule deposits per se. This line of reasoning also suggests that there is no persuasive reason to interfere with market forces in determining the rate of exploitation of manganese nodules. To do so would be to jeopardize the achievement of allocative efficiency, the topmost (in some cases the only) value posited by this approach.[42] Moreover, an open-to-entry regime for deep-seabed mining would avoid both the transaction costs and the dangers of economic distortions that are associated with the operation of supranational institutions. In sum, those advocating this view assert that supranational institutions are unnecessary for the achievement of efficient outcomes in deep-seabed mining and that the introduction of some such regime might generate large, avoidable costs.

Yet this line of reasoning has serious flaws. There is a pervasive feeling among potential investors that exclusive rights to manganese nodules and security of tenure are necessary to call forth large-scale investment in deep-seabed mining. An officer of the Chase Manhattan Bank, for example, has testified that "a firm concession with security of tenure to a specific mine site would appear to me to be an absolute requirement for the project financing of an underseas mining venture—without it, a lender could assume neither the reserve nor production risks."[43] And most would concur with Christy's conclusions to the effect that "*economic efficiency* in production will require, first of all, the ability of the entrepreneur to acquire security of investment. He must be able to obtain exclusive rights to a sufficiently large area for a sufficient length of time to capture a fair return on his investment. While it is conceivable that an entrepreneur could undertake such an investment and operate it effectively in the absence of rules guaranteeing tenure of rights, most writers believe that the risk would be too high to do so."[44]

The argument for an open-to-entry regime for deep-seabed mining also fails to consider issues other than efficiency (defined in strictly economic terms). It makes no provision, for example, for efforts to

41. Eckert, "Exploration of Deep Ocean Minerals," 163.

42. Thus, Eckert states explicitly that "the general criterion will be the efficient allocation of scarce resources—including nodules and the allied resources required to gather and process them" (ibid., 144).

43. U.S. Congress, Committee on Merchant Marine and Fisheries, *Report on the Deep Seabed Hard Minerals Act,* Report No. 95–583 (1977), 21.

44. Francis T. Christy, "A Social Scientist Writes on Economic Criteria for Rules Governing Exploitation of Deep Sea Minerals," *International Lawyer* 2 (1968), 229; emphasis in original.

control externalities of deep-seabed mining that affect the quality of the marine environment. Nor does it allow for the pursuit of political goals such as the protection of existing copper and nickel operations located in developing countries or the transfer of certain types of technology to developing countries. Similarly, an open-to-entry regime for deep-seabed mining would raise serious distributive questions. Initially, all economic returns (including economic rents) would fall into the hands of the mining companies under a regime of this sort. It is true, of course, that this outcome could be altered through interventions on the part of governments and the negotiation of agreements covering suitable international transfer payments.[45] Yet there is little reason to expect the governments of the relevant countries to pursue such a policy vigorously, a fact that makes it easy to understand the objections of the less developed countries to this approach to the distributive issues associated with deep-seabed mining.

Perhaps even more decisive is the recognition that the case for an open-to-entry regime for deep-seabed mining rests on a mistaken conception of existing rights to the manganese nodules of the deep seabed. The deposits of nodules are in fact currently owned; they are not literally unowned or res nullius. This conclusion is inherent in the proposition that marine resources are common property, and it is reaffirmed explicitly in United Nations General Assembly Resolution 2749 (XXV) of 1971, which specifically describes deposits of nodules as forming part of the "common heritage of mankind" (a resolution voted for by the United States and other developed states).[46] As it happens, the owner of the nodules is the international community as a whole rather than certain individual members of this community. But this surely does not imply an absence of ownership or of property rights any more than public ownership does in municipal systems. No doubt, the international community could decide to allow individual members to exploit the nodules without restrictions in the forms of rules governing use or the payment of fees.[47] But it need not do so. Certainly, the owner of a given natural resource has the right to specify the conditions under which the resource will be available for exploitation. In the case at hand, the owner (that is, the international community) has expressed quite

45. Committee on Merchant Marine and Fisheries, *Deep Seabed Hard Minerals Act,* 39.
46. Consult the discussion of Resolution 2749 in Michael Hardy, "The Implications of Alternative Solutions for Regulating the Exploitation of Seabed Minerals," *International Organization* 31 (1977), 313–315.
47. This is the traditional position of the international community regarding marine fisheries. But even in the fisheries, there can be no question that the international community has the authority to impose rules restricting use, if it should choose to do so.

clearly its desires concerning the use of the resource. Thus, Resolution 2749 requires the establishment of "appropriate international machinery" to govern the exploitation of the manganese nodules. It lays down goals to be pursued in this connection (for example, paying "particular consideration [to] the interests and needs of the developing countries"). It specifies that the details of the regime for deep-seabed mining are to be incorporated into an "international treaty of universal character." And it clearly implies that the individual members of international society are not free to proceed with the exploitation of the nodule deposits unilaterally, pending the establishment of the relevant supranational institutions. Under the circumstances, the fact that disagreements over the details of this regime have delayed the completion and ratification of the required treaty can hardly justify individual members' taking matters into their own hands and proceeding to exploit the nodules unilaterally.[48] To do so would be roughly equivalent to mining firms simply going ahead with the extraction of hard minerals on publicly owned lands when there is disagreement within the relevant community over the formulation of specific rules governing the exploitation of the minerals.

Assume, then, that some supranational regime for deep-seabed mining will arise, whether or not these arrangements meet with universal approval. What could or should such a regime look like? It turns out, of course, that numerous variations are possible and that the choice of any one will have significant consequences for the parties involved. To set the stage for a discussion of the principal institutional options I shall spell out some key distinctions at the outset.

Any of the options likely to seem adequate to handle deep-seabed mining will require some administrative apparatus or set of organizations. For convenience, label this apparatus the International Seabed Authority (following the usage of the Third United Nations Conference on the Law of the Sea and the 1982 Convention on the Law of the Sea).[49] Would the International Seabed Authority (ISA) actually own the manganese nodules of the deep seabed or would it merely be vested with managerial authority by the international community?[50] If the ISA is merely a manager or a steward, what obligations will it have to the actual

48. For a conflicting view see Committee on Merchant Marine and Fisheries, *Deep Seabed Hard Minerals Act*, 24–25.

49. Third United Nations Conference on the Law of the Sea, *United Nations Convention on the Law of the Sea*, A/Conf.62/122 (New York), October 1982.

50. For general background regarding structures of property rights consult Dales, *Pollution, Property, and Prices*, and Furubotn and Pejovich, "Property and Economic Theory."

owner of the nodules (the international community), and in what ways would the community be able to intervene to review or revise the regime? In fact, the community is not likely to transfer formal property rights in the nodules to the ISA. Yet the community is also unlikely to interfere with its day-to-day operations because of the unwieldy character of the international community and the consequent difficulty of reaching collective decisions at the international level. Additionally, there is the question of whether the ISA should become an operating authority (that is, what is called the Enterprise in the 1982 Convention) or remain a regulatory authority, supervising (perhaps closely) the activities of the private companies or state corporations that undertake the actual work of extracting, transporting, and processing the manganese nodules.[51] The basic issue here is the extent to which deep-seabed mining is to proceed in the fashion of a regulated private-enterprise system or in the fashion of a socialized system operating through some international analog of a state corporation.[52] Current discussions generally emphasize a mixed system in which some combination of these options is pursued. In fact, it seems likely that little deep-seabed mining undertaken during the foreseeable future will be carried out by the Enterprise. By the same token, however, there can be no doubt that the ISA (if and when it actually begins to function) will want to supervise closely the activities of the corporations or corporate consortia engaged in deep-seabed mining operations.

Rate of Development

What is the appropriate rate of extraction of the manganese nodules? Should the ISA have the authority to make binding decisions about this issue? If so, how actively should it intervene to influence rates of harvest of manganese nodules? Since the nodules do not constitute a renewable resource (within any relevant time frame), there is no need to worry about maintaining healthy stocks. At the same time, it appears that existing stocks (that is, inventories) of nodules are so extensive that they are likely to be effectively inexhaustible within any pertinent time frame.[53] Worries about the supply literally running out, therefore, are

51. Scc Hardy, "Regulating the Exploitation of Seabed Minerals," 321–328.
52. For a theoretical discussion of the less familiar, socialized alternative see Oscar Lange and Fred M. Taylor, *On the Economic Theory of Socialism*, ed. Benjamin E. Lippincott (Minneapolis: University of Minnesota Press, 1938).
53. Allen L. Hammond, "Manganese Nodules I," *Science* 183 (8 February 1974), 502–503. Of course, this fact does not mean that all (or even a large fraction) of the nodules can be exploited profitably under current conditions with respect to technology and world market prices.

less meaningful in the case of manganese nodules than in the case of other nonrenewable resources such as oil and natural gas. Under the circumstances, the issue at stake here boils down to the questions of whether there is some socially desirable rate of use for manganese nodules and whether the unregulated activities of corporate actors can be expected to approximate this rate.

As it turns out, however, these questions raise a profound problem of managerial philosophy. What do we wish to maximize in deep-seabed mining? Are we content to let the markets for the relevant private goods (that is, copper, nickel, cobalt, and managanese) operate in an unregulated fashion so that the rate of use of managanese nodules is determined by the interplay of supply and demand in these markets? Or do we wish to apply various nonmarket or "political" criteria to replace or supplement these market considerations in settling on the rate of use for manganese nodules? Not surprisingly, there is no shortage of proposals for nonmarket criteria. The 1982 convention, for example, requires the ISA to regulate the production of nodules in such a way as to ensure "the protection of developing countries from adverse effects on their economies or on their export earnings resulting from a reduction in the price of an affected mineral . . . to the extent that such a reduction is caused by activities" in the deep seabed (Article 150). This and other proposals like it are protectionist devices designed to safeguard the financial health of competing mineral producers already operating in the less developed countries (or elsewhere?). Similarly, there are those who wish to restrict the exploitation of manganese nodules to minimize social costs that will not be reflected in market prices. The critical issue here is the extent to which deep-seabed mining is likely to generate negative externalities affecting the quality of the marine environment.[54] On the other hand, some commentators have argued that the rate of use of manganese nodules resulting from market forces will be too low and steps should be taken to accelerate this rate. There is some reason to believe that the industries involved in deep-seabed mining will be less than fully competitive so that there may be a need to combat production restrictions designed to reap monopoly profits or economic rents. Even more to the point, some American policymakers wish to accelerate the production of minerals from nodules in the interests of achieving what amounts to independence in strategic minerals for the United States.[55] The argu-

54. This subject has been examined, in a preliminary way, in the Deep Ocean Mining Environmental Study (DOMES) carried out by the National Oceanic and Atmospheric Administration (NOAA). For a discussion of the results consult Office of the Comptroller General, *Deep Ocean Mining Environmental Study*.

55. See also Committee on Merchant Marine and Fisheries, *Deep Seabed Hard Minerals Act*, 17–19.

ments for this idea bear a distinct resemblance to the arguments for the parallel notion of achieving energy independence.

No doubt there are other nonmarket criteria that could be applied in discussions of a socially optimal rate of use of manganese nodules. In general, however, it is important to recognize that the application of any nonmarket criterion will necessitate some political choice and that the transaction costs associated with such choices are often high. This is the chief virtue of relying on markets—they yield collective choices while minimizing the transaction costs of doing so. I do not find this virtue compelling, at least when it comes to making decisions on rates of use of manganese nodules. Though this is admittedly an area requiring value judgments, there are good reasons to be concerned about protecting the marine environment and safeguarding certain industrial enterprises in the less developed countries. In any case, I can see no persuasive reason to deny the ISA the authority to make binding decisions on the rate of use of manganese nodules. To do so would be to settle in advance ideological issues over which there are fundamental and legitimate differences of opinion.

Conditions of Entry

On what terms should corporations or consortia be granted access to manganese nodules? The issue here concerns the general rules under which deep-seabed mining takes place rather than the allocation of shares or tracts among those desiring to extract nodules. Were the ISA to become an operating authority (that is, rely exclusively on the Enterprise), this issue would become irrelevant. As I have already argued, however, this is an improbable scenario. At the opposite extreme, there is the argument, discussed earlier in this section, that an open-to-entry regime with no rules governing access to the nodules would be perfectly adequate for deep-seabed mining. Such a system would permit individual firms to dredge for nodules at places and times of their own choosing, without regard to the concurrent efforts of others or standards intended to promote an orderly pattern of mining development. But the firms themselves are likely to reject this option, quite apart from the social costs that an arrangement of this sort would entail. The issue of rules governing access to manganese nodules is therefore a serious one deserving careful consideration in the process of devising a regime for deep-seabed mining.

There are at least five distinct questions in this realm. The first relates to the identification of mining sites or nodule tracts. What size should tracts be? How should they be demarcated? Who should determine what tracts to make available during any given time period? These are largely technical questions.[56] While they are far from trivial, there is no reason for them to become a focus of severe conflict. The same cannot be said of a second concern: should firms interested in deep-seabed mining be granted guaranteed access to nodule deposits?[57] It is generally accepted that the ISA would have the authority to issue mining licenses, and it seems reasonable to require that firms acquire licenses before they undertake dredging operations. The real questions concern the criteria and rules governing the issuance of such licenses. Should firms be guaranteed the receipt of licenses upon filing the required applications? Is there any way to ensure that licensing procedures are not abused in the pursuit of political interests? If the ISA is to manage deep-seabed mining effectively, it must surely possess the authority to grant licenses and establish procedures for the issuance of these licenses. This undoubtedly means that the possibility of political abuse cannot be altogether eliminated. But there is nothing unusual or remarkable about this.

More substantively, there are three questions regarding the rights accorded to firms under ISA licenses. Should individual firms be granted exclusive rights to dredge within the boundaries of specific tracts? There is, as indicated previously, some debate about the economic significance of exclusive rights in deep-seabed mining. But there is little doubt that firms will demand exclusive rights to dredge on specific tracts, and it is hard to see any reason for the ISA to refuse to accommodate such demands. Much the same can be said about the issue of security of tenure. Should individual firms be guaranteed exclusive rights to dredge on specific tracts *over time*? There is a case for placing some time limits on the duration of licenses (ten years, twenty years?), requiring some economic activity in order to retain a claim, and demanding that holders of claims comply with the general rules formulated by the ISA.[58] But beyond this, security of tenure is a reasonable demand on the part of mining firms, and it is hard to see why the ISA would oppose it. A very different issue arises in connection with investment guarantees. Is it desirable to create a program designed to guarantee or protect the investments of private firms engaged in deep-seabed mining? In my

56. For a discussion of these questions see Hardy, "Regulating the Exploitation of Seabed Minerals," 315–321. On the treatment of similar issues in connection with outer continental-shelf oil and gas development see U.S. Senate, Committee on Energy and Natural Resources, *Report on the Outer Continental Shelf Lands Act Amendments*, Report No. 95–248 (1977), 48 et seq.

57. Samuelson, "Law of the Sea Treaty," 1342.

58. Article 153 of the 1982 convention, for example, addresses these issues by requiring operators to carry out their activities under the terms of a "formal written plan of work."

judgment, such guarantees could only be justified in situations characterized by an absence of exclusive rights and security of tenure. Otherwise, guarantees of this type would appear to be a political response to
the special concerns of a powerful interest group.[59]

Allocation of Shares

How should shares in manganese nodules be apportioned when there
is competition among operators desiring access to them? The problem
here boils down to an examination of alternative methods of assigning
rights (or licenses conveying rights) in specific mining sites or nodule
tracts. Assuming that the ISA has already made decisions about the
number and the identity of tracts to be made available during any given
time period as well as about the conditions of access, it remains only to
devise a procedure for apportioning exclusive rights in individual tracts
among those desiring to engage in mining. But this turns out to be a
complex issue in its own right.[60] There are several distinct procedures
for apportioning such rights, each of which has some drawbacks under
the conditions likely to prevail.

There is, to begin with, the option of proceeding on a first come, first
served basis. The first firm to stake a claim to a given nodule tract would
receive priority in filing for a license to dredge on that tract. To ensure
exclusivity of rights under this procedure, the ISA would have to record
or even patent individual claims. But this is no more than a technical
problem. This option would resemble the American procedure for handling claims to land-based minerals under the provisions of the Mining
Act of 1872.[61] It would also have the drawbacks of this procedure.
Serious inefficiencies would probably occur in the forms of stampedes to
stake extensive claims and overcapitalization. Nor is there any reason to
expect that the resultant distribution of claims would conform to even
the most modest standards of equity. Additionally, it may help to distinguish two variants of this procedure. The ISA could register claims
without charging fees (or levying only nominal fees to cover administrative expenses), in which case the nodules themselves would be free
goods from the perspective of the mining firms and there would be no

59. Such guarantees are also envisioned under Title II of the U.S. Deep Seabed Hard
Minerals Resources Act of 1980 (PL 96–283, 30 USC 1401 et seq.) to compensate for losses
caused by the entry into force of an international treaty.

60. The pros and cons of various options are surveyed in connection with outer continental-shelf oil and gas development in Devanney, *OCS Petroleum Pie*, 68–118.

61. An Act to Promote the Development of the Mining Resources of the United States,
U.S. Statutes at Large, 1871–1973, 91–96.

economic returns to the owner of the nodules. Alternatively, the ISA could levy nontrivial fees or charges in conjunction with the issuance of licenses; this procedure would raise interesting questions about revenues, which are discussed later in this section.

A second procedure for allocating shares features administrative decisions coupled with the issuance of work-obligation permits. Under this option, firms interested in specific nodule tracts would submit proposed plans for exploitation to the ISA, which would then select the "best" of the proposals and grant licenses or work-obligation permits on this basis. The problems afflicting this procedure, however, are severe: individual firms have incentives to submit unrealistic plans; there is scope for political abuse in the administrative decision process; it is not easy to secure compliance with the terms of work-obligation permits, and the transaction costs of the system are apt to be high.[62] While it is true that the procedure facilitates serious consideration of nonmarket concerns such as maintaining the quality of the marine environment, it is doubtful whether this advantage is sufficient to offset the drawbacks intrinsic in the procedure.

Various bargaining processes could also be employed in apportioning nodule tracts among mining firms. One idea is simply to allow firms to strike private bargains (formalized in enforceable contracts) in which they would agree among themselves about who is to exercise exclusive rights to specific tracts.[63] Such bargaining may not yield mutually acceptable outcomes, and of course there is no guarantee that any outcomes reached will be equitable. Quite apart from these concerns, however, this procedure would bypass the ISA with respect to the issue of apportioning nodule tracts among users. It would therefore seem inappropriate for the type of regime for deep-seabed mining under consideration here. To address this concern, individual firms might be allowed to engage in bilateral bargaining with the ISA over the issuance of licenses for specific nodule tracts. Such a procedure would amount to a kind of intermediate system between work obligation permits and full-fledged competitive bidding (see below). In practice, this procedure might well exhibit most of the drawbacks of a system of work-obligation permits. The scope for political abuse would be substantial so that serious problems of equity would arise. Additionally, the transaction costs of this system would be high, especially if the ISA allowed several firms to become involved in individual bargaining sessions in the interests of reducing complaints about unfairness to a minimum.

62. Devanney, *OCS Petroleum Pie*, 80–82.
63. Eckert, "Exploration of Deep Ocean Minerals," 157–161.

Turn, then, to the option of creating a market or a quasi market in rights to exploit individual nodule tracts.[64] Here the ISA would periodically select a group of tracts to be made available for exploitation and announce a license (or lease) sale at a specified time. Exclusive rights to dredge on a given tract would then go to the firm making the highest bid at the time of the sale.[65] It turns out that there are numerous variants on this type of procedure. Recent debates over the allocation of oil and gas tracts on the American outer continental shelf, for example, have been full of acrimonious arguments about the alleged advantages and disadvantages of the various methods of organizing lease sales.[66] The basic problems associated with this option have to do with maintaining a reasonable level of competition in the bidding and providing real opportunities for smaller firms or less developed countries to play some role in deep-seabed mining. Yet the idea of a market in exploitation rights for individual nodule tracts has significant attractions: it could provide the ISA with sizable revenues, and the transaction costs of operating the system need not be excessive. The ISA might want to give serious consideration to carrying out preliminary exploration of nodule tracts on its own before putting the relevant exploitation rights on the market (compare the analogous debate regarding exploratory work in conjunction with the leasing of oil and gas tracts on the outer continental shelves by the U.S. government).[67] Also, the ISA would undoubtedly want to incorporate relatively stringent restrictions into the licenses issued for nodule tracts to control social costs, especially those affecting the quality of the marine environment. Nonetheless, there is a strong case to be made for properly organized markets in exploitation rights as the preferred procedure for allocating shares of manganese nodules.

Nonmarket Effects

Though a market system may have virtues once the ISA decides to make specific nodule tracts available for current use, this conclusion should not preclude the ISA from devoting attention to nonmarket effects in the management of deep-seabed mining. In fact, a few examples will demonstrate that the ISA could easily emphasize a variety of

64. See Dales, *Pollution, Property, and Prices*, chap. 6 for a comparable discussion of pollution rights.
65. A similar system is currently used in the United States to allocate leases to oil and gas tracts on the outer continental shelf. See Devanney, *OCS Petroleum Pie*, 68–79.
66. See Committee on Energy and Natural Resources, *Outer Continental Shelf Lands Act Amendments*, especially 158–161.
67. Devanney, *OCS Petroleum Pie*, 96–118.

nonmarket concerns, even while relying on competitive markets in exploitation rights to individual nodule tracts.

Consider first the issue of social costs. The social costs of greatest concern are those pertaining to the quality of the marine environment, though other social costs might arise as well. Though there is some debate about the probable environmental impacts of deep-seabed mining, it is reasonable to believe that the environmental problems will be less severe than those associated with offshore oil and gas operations.[68] Even so, the ISA ought to have the authority to make independent assessments of such matters and to take effective steps to control social costs. This may require, in some cases, withholding specific nodule tracts from exploitation, regardless of the desires of the mining companies. In other cases, the promulgation of well-defined standards coupled with a system of charges should suffice to control the pertinent threats to environmental quality.[69]

Additionally, it is hard to see any inherent objection to efforts on the part of the ISA to achieve political goals in managing deep-seabed mining.[70] It may seem desirable, for example, to withhold certain tracts that are rich in nodules from current use in the hope that firms from the developing countries will be in a position to bid on them at a later date. Similarly, the ISA may wish to withhold some tracts for its own use in the event that it decides to pursue the idea of activating the Enterprise. There have also been suggestions that the ISA should seek to promote technology transfers by compelling mining companies to disclose information about the technology employed in their mining operations.[71] Many efforts along these lines will undoubtedly be opposed on the grounds that they promote economic inefficiencies. Still, it is important to note that the conception of economic efficiency underlying most

68. Since the mining operations will take place in the deep ocean, the resultant environmental problems are not likely to be of great concern to fishing interests or to resort operators worried about the preservation of beaches. Still, see the comments in Committee on Merchant Marine and Fisheries, *Deep Seabed Hard Minerals Act,* 35, to the effect that "there is a real concern about the possible adverse impact which deep seabed mining may have on the environment. But this concern is tempered by the generally widespread recognition that these impacts are, at this point, unpredictable."

69. On charges, see Allen V. Kneese and Blair T. Bower, "Standards, Charges, and Equity," 217–228 in Dorfman and Dorfman, *Economics of the Environment.*

70. My prior criticism of the FCMA regime on grounds of politicization should not be confused with this argument about the legitimacy of considering political goals. In the case of the FCMA, the problem is that the decision-making apparatus permits special interest groups with no particular claim to legitimacy to influence outcomes decisively. Here the point is that there is no inherent reason for the relevant community to eschew political goals in managing deep-seabed mining.

71. Some specific provisions regarding technology transfer are included in Article 144 of the 1982 convention.

criticisms of this type is a rather narrow one that rests on the idea of achieving an outcome lying on a production-possibility frontier defined largely in terms of standard private goods. While efforts to achieve specific nonmarket objectives should no doubt be scrutinized critically, the fact that they interfere with the pursuit of economic efficiency is surely not a definitive objection.

This discussion suggests one additional comment about the design of a deep-seabed mining regime. It seems clear that the ISA should have legal personality, in the same way that government agencies do in municipal systems. Such a status would be important not only to enable the ISA to bring suits but also to allow others to file suits against it. It would permit the ISA to employ legal sanctions against mining companies that violate the terms of their licenses, fail to pay charges levied in conjunction with environmental standards, or refuse to disclose information about mining technologies. At the same time, it would allow mining companies to sue the ISA for alleged violations of their entitlements under the terms of mining licenses and outside groups to sue the ISA on the grounds that the agency had failed to enforce its own rules and standards vigorously. The principal complication would be that the International Court of Justice is not capable of handling the relevant litigation.[72] As a result, it would help if the ISA could obtain legal standing in the courts of individual members of the deep-seabed regime.[73]

Revenues

Should the ISA manage the extraction of manganese nodules in such a way as to generate revenues? There are at least three distinct issues embedded in this question. There is, to begin with, the issue of whether the ISA should levy user's fees on mining firms to cover its own administrative expenses. Surely, the ISA should be granted this authority, and it would be easy enough to handle the problem through the device of licensing fees. The next issue concerns the extent to which the ISA, acting as the agent of the actual owner of the manganese nodules, should endeavor to recover normal economic returns from the nodules regarded as a valuable factor of production. A private owner of a natural resource would act in this way, and there is apparently some trend in this

72. For a study that clearly substantiates this conclusion even though it is sympathetic to the Court see Shabtai Rossenne, *The World Court*, 3d rev. ed. (Dobbs Ferry, N.Y.: Oceana, 1973).

73. The ISA might also consider creating a judicial mechanism of its own along the lines of the Court of Justice of the European Community.

direction at the level of individual states. Most governments today, for example, seek to collect normal economic returns from deposits of oil and natural gas under their jurisdiction.[74] The idea of proceeding in this fashion at the international level has been hampered by the absence of strong organizations as well as by the operation of the "law of capture" traditionally associated with common-property regimes. The marine fisheries undoubtedly are a dramatic case in point. Yet the whole idea of creating supranational institutions to manage deep-seabed mining constitutes a break with past practices in the international relations of resource management. Accordingly, it is hard to see any compelling reason why the ISA, acting on behalf of the international community, should shy away from collecting normal economic returns on manganese nodules extracted by the mining companies. Beyond this, we might ask whether the ISA should seek to collect a share of any true economic rents resulting from mining (that is, scarcity rents or surplus returns to factors of production that arise from market imperfections).[75] Admittedly, it is impossible to avoid value judgments in this area, and it might well be politically impossible for the ISA to succeed in attempts to collect true economic rents. Yet the case for the ISA benefitting from such rents is analytically equivalent to the argument for transferring economic rents to the public sector in domestic systems.[76]

If the idea of allowing the ISA to raise revenues is accepted, it becomes important to consider procedures for obtaining such revenues. Not surprisingly, numerous procedures are available: cash sales, bonus bids, license fees, royalties, taxes of one sort or another, or combinations of these devices. The choice of specific mechanisms is a technical matter requiring expert analysis. But the basic principle that the ISA should follow is simple enough: the Authority should endeavor to collect its share of the proceeds from deep-seabed mining in such a way as to distort the economic calculations of the mining firms as little as possible. In the case of normal economic returns, a straightforward system of royalties (computed as a percentage of market price) might well suffice. After all, normal returns are (or should be) part of any economic enterprise; the activity should not be undertaken at all if such returns are not forthcoming.[77] The issue of true economic rents is more complex. My

74. Interestingly, many governments still do not seem to think in terms of collecting normal economic returns from publicly owned or managed fisheries and forests.

75. See also Helen Hughes, "Economic Rents, the Distribution of Gains from Mineral Exploitation, and Mineral Development Policy," *World Development* 3 (1975), 811–825.

76. See also Devanney, *OCS Petroleum Pie*, in which the issue is discussed in terms of the distinction between national income and public income.

77. This is the system currently in place in the United States with respect to oil. See also Resources for the Future, *U.S. Energy Policies: An Agenda for Research* (Washington, D.C.: Resources for the Future, 1968).

feeling is that there is a strong case for something like an excess-profits tax, but this is not the place to present an extensive discussion of this matter.

To the extent that the ISA accumulates revenues in excess of its administrative expenses (and this would surely occur given the suggestions outlined above), it would be necessary to come to terms with the distribution of these funds. There are at least three distinct options. The extra funds could be set aside or placed in escrow to be used during international emergencies at the discretion of the ISA itself or some other international body, such as the United Nations.[78] Alternatively, the funds could be distributed to the "shareholders" of the ISA (that is, the individual members of the international community). In effect, this would amount to dividing the funds among the signatories to the deep-seabed mining treaty according to some agreed-upon formula (for example, on the basis of population). Or the funds could be used to pursue various redistributive goals through the mechanism of some system of international transfer payments. Though numerous goals of this sort come readily to mind, it seems relevant at this point to recall the injunction of Resolution 2749 to pay particular attention to the "interests and needs of the developing countries." Once again, we have reached a point where value judgments are required. In my view, some combination of the first and third alternatives would seem attractive in dealing with this distributive question. While the case for planned redistribution is strong, there is much to be said for setting up at least a modest international emergency fund to be used for disaster relief on a worldwide basis.

Costs

It seems reasonable, then, to conclude that the ISA would be able not only to cover its own expenses but also to accumulate surplus revenues. But what about the costs to international society of establishing and operating supranational institutions to manage deep-seabed mining? What would be the magnitude of these costs, and would the benefits accruing to international society from such a regime exceed these costs? Several types of costs require consideration in any attempt to respond to these questions.

At the outset, there would be start-up costs associated with setting up the regime and, especially, the ISA. These would be substantial if the ISA were to become an operating agency that engaged in deep-seabed

---

78. The issue of appropriate decision-making procedures for the allocation of such funds would no doubt be controversial. There is much to said, however, on grounds of efficiency and equity for a specialized agency staffed by professionals in contrast to the United Nations per se.

mining in its own right; initial capital investments to organize an efficient operation would run into the hundreds of millions of dollars.[79] If the ISA were to act essentially as a regulatory agency, however, start-up costs should be more modest. The largest items would be a physical facility and various types of equipment, and the costs of these could hardly assume exorbitant proportions. Once the ISA was underway, there would also be operating expenses, including decision costs, administrative costs, and compliance costs. We might group these loosely under the broader rubric of transaction costs.[80] Though these transaction costs would certainly not be trivial, it is hard to make confident estimates of their magnitude, which would depend on such things as the intensity of conflicts of interest among the members of the regime, the willingness of individual members to accept the use of various procedural devices to resolve conflicts of interest, and the propensity of the mining companies themselves to comply with the decisions of the ISA. There can be little doubt, moreover, that these costs would rise as a function of the number of members in the regime, the degree to which individual members were accorded anything like a veto power in the decision processes, and the extent to which the ISA pursued political goals in managing deep-seabed mining. While the specification of a concrete figure would be arbitrary, my judgment is that the operating or running costs of the regime would be substantial.

Additionally, the regulatory efforts of the ISA might prove costly in the sense of interfering with the economic efficiency of deep-seabed mining. As Eckert observes, "Regulation could harm economic efficiency by raising the costs of entry into ocean mining, limiting output, and introducing monopoly to protect terrestrial producers of metals."[81] What is more, the ISA might encounter so much difficulty in reaching clear-cut decisions that it would severely curtail the pace of deep-seabed mining, thereby keeping investments below what would be required for the achievement of economic efficiency.

Though inefficiencies of this sort would certainly be possible in a deep-seabed mining regime, they would not be necessary. Certainly the ISA could pursue the goal of economic efficiency in a single-minded manner if the members chose to do so. What is more, the achievement of economic efficiency would hardly be guaranteed in the absence of a regime featuring supranational institutions. The general view, in fact, is that an open-to-entry regime would encourage relatively serious forms

79. Hardy, "Regulating the Exploitation of Seabed Minerals," 316.
80. See also Sandler and Cauley, "Design of Supranational Structures."
81. Eckert, "Exploration of Deep Ocean Minerals," 174–175.

of economic inefficiency. There are many who would prefer some risk of underinvestment as a result of the activities of the ISA to the risk of overinvestment under an open-to-entry regime. It is important to remember, moreover, that the costs of these economic distortions would often be offset by benefits from the ISA's efforts to achieve political goals (for example, providing developing countries with advanced technologies).[82] Whether the resultant political benefits would outweigh the economic costs is hard to say; conclusions about such matters generally rely heavily on value judgments. As the discussion earlier in this section suggests, however, there is nothing inherently objectionable in policies that require certain economic sacrifices to achieve political goals.

This analysis leads to two additional observations regarding costs. It is not possible to arrive at any simple answer to the question of whether the total benefits to international society from such a regime would exceed the total costs. This problem arises, in part, because it is difficult to conduct a controlled experiment, one in which some deep-seabed mining activities were managed through a regulatory arrangement and others were left to an open-to-entry regime. Even more important, the problem stems from the lack of a usable metric for computing all the costs and benefits flowing from the creation of supranational arrangements to manage deep-seabed mining. This will continue to be the case unless and until the difficulties associated with the construction of social-welfare functions are solved.[83] In my own judgment, however, the political advantages likely to flow from the operation of the ISA might well be sufficient to outweigh any economic costs associated with it. Finally, it is worth emphasizing that it should be possible to avoid severe conflict over the *incidence* of the costs of operating the ISA. This is a consequence of the fact that the ISA would be able to cover its actual expenses from its own revenues and would not be compelled to raise revenues through taxation or some similar procedure to sustain itself. This condition does not constitute an excuse to treat lightly the costs of operating the ISA. But it will eliminate the sort of contention over scales of contributions that has plagued other international organizations, such as the United Nations.[84]

---

82. Another way of making this point is to say that Pareto optimality does not necessarily require choosing an outcome on the production-possibility frontier, so long as this frontier includes only private goods exchanged through market mechanisms.

83. For a clear discussion of the problems in constructing social-welfare functions see Dorfman and Dorfman, *Economics of the Environment*, 11–13.

84. See John G. Stoessinger and Associates, *Financing the United Nations System* (Washington, D.C.: Brookings, 1964).

Compliance

Turn, finally, to the problem of compliance with the rules of a deep-seabed mining regime and the specific decisions of the ISA. What is the likelihood that relevant actors would comply with these rules and decisions? What recourse would the ISA have in dealing with violators or those that fail to show clear evidence of compliance? Surely, this is an important issue. It is not hard to imagine the American government lending support to noncompliant behavior on the part of its nationals if the ISA were to make decisions conflicting with the American position on issues such as guaranteed access to nodule tracts.[85] Nor is it necessary to single out the United States for special criticism. Some governments would undoubtedly exhibit even fewer scruples.

Accordingly, the ISA would face two distinct questions.[86] It could choose to allocate some of its resources to programs designed to elicit compliance with its own decisions and the rules of the regime. It is easy to assert that the ISA should invest in such programs until the marginal gains (measured in terms of the benefits attributable to a reduction in violations) just equal the marginal costs. So stated, however, this is hardly an operational criterion. The development of such a criterion would require an analysis going well beyond the scope of this chapter. But as a preliminary judgment, it seems likely that the ISA would want to invest substantial resources in the achievement of compliance.[87]

The other question concerns methods. Assuming that the ISA does make some investment in the achievement of compliance, what sorts of compliance mechanisms would it find most cost effective? It seems clear that physical coercion is not a viable option for the ISA. The members of the regime are hardly likely to allow the ISA to equip itself with a police force. Nor is it likely that a standby force composed of contingents from individual members would prove any more workable in this case than under the auspices of the United Nations.[88] Rather, the ISA will probably find it expedient to employ nonphysical measures and to concentrate directly on the behavior of the mining firms. Thus, it could require each firm to set up an escrow account prior to the receipt of licenses to be used

85. Samuelson, "Law of the Sea Treaty."
86. For a theoretical examination of the general problem of compliance see Oran R. Young, *Compliance and Public Authority: A Theory with International Applications* (Baltimore: Johns Hopkins University Press, 1979).
87. This judgment rests more on the premise that mining firms would experience strong incentives to engage in certain violations than on any detailed assessment of the costs of eliciting compliance in the realm of deep-seabed mining.
88. Lincoln P. Bloomfield, ed., *International Military Forces* (Boston: Little, Brown, 1964).

to cover potential fines or damages. Firms violating the rules of the regime could be prohibited from participating in subsequent license sales for some period of time. Alternatively, firms compiling favorable records of compliance could be given various breaks regarding the payment of fees or royalties. The ISA could greatly simplify the compliance problem, of course, by persuading individual states to allow it to bring legal action against mining companies through municipal courts. Such arrangements would not only allow the ISA to pursue violators in an orderly fashion, they would also facilitate enforcement actions in specific cases. The response of individual states to this idea is difficult to gauge at present. The American reaction would certainly be critical, since a high proportion of the firms engaged in deep-seabed mining would be corporations chartered under the laws of the United States.[89] While some such arrangement might be supported by the Department of State, powerful interests in the business and financial communities would almost certainly oppose it. The ultimate American reaction would therefore turn on complex maneuvers in the political system of the United States. Yet this discussion strongly reconfirms the point made earlier in this section about the advantages to the ISA of achieving legal standing in the municipal courts of the individual members of international society.

CONCLUSION

The study of comparative statics is, in essence, a matter of assessing the performance of institutional arrangements in terms of well-defined criteria of evaluation. Commentators may disagree about the appropriate criteria to use in such endeavors. Though economists typically fix on allocative efficiency as their topmost consideration, for example, many others see nothing compelling about this ranking. Some criteria of evaluation are notoriously difficult to define precisely. The criterion of equity is certainly a leading case in point. Moreover, it is often hard to determine the probable performance of institutional arrangements in terms of various criteria, especially when we focus on real-world cases (for example, actual markets) in contrast to ideal types (for example, perfect competition markets). Nonetheless, the fundamental character of comparative statics as an analytic exercise is clear.

Conclusions regarding the attractions of regional arrangements for

89. Eckert, "Exploration of Deep Ocean Minerals," 144–151, and Samuelson, "Law of the Sea Treaty," 1339.

the fisheries or specific licensing arrangements for deep-seabed mining, however, tell us nothing about the processes of regime formation that give rise to the institutional arrangements that actually emerge in international society. To the dismay of those whose work emphasizes comparative statics, processes of regime formation at the international level regularly yield results that diverge sharply from the arrangements judged optimal by outside analysts. While the study of comparative statics is surely important in providing us with benchmarks against which to measure the performance of actual regimes, therefore, it is critical to turn now to an examination of the processes of regime formation in international society. The next two chapters take up this challenge.

CHAPTER SIX

Regime Formation as Contract Negotiation: Nuclear Accidents

Nuclear accidents have become a fact of life. Though we cannot predict when or where or how frequently nuclear accidents will take place, we can be certain that they will happen from time to time. Given the nature of nuclear technology and the behavior of radioactive materials, many of these accidents will have transboundary consequences affecting natural environments and human communities hundreds or even thousands of miles from the actual sites of the accidents. The more serious nuclear accidents, like the 1986 Chernobyl disaster, will produce radioactive fallout measureable on a global scale and capable of contaminating large ecosystems for decades. These observations suggest the importance of creating a fully developed international regime to deal in an orderly fashion with an array of issues associated with nuclear accidents.

This chapter advances and examines the proposition that the nuclear-accident problem, by contrast with other transboundary environmental problems such as acid precipitation, carbon dioxide buildup, and even ozone depletion, has several characteristics that make it an attractive focus for those interested in the development of institutional arrangements in international society. Already, over fifty states (including the United States and the Soviet Union), negotiating under the auspices of the International Atomic Energy Agency, have signed agreements to establish early warning mechanisms and emergency relief systems for nuclear accidents.[1] Following a discussion of incentives to enter into a nuclear-accident regime, the chapter considers what a fully developed

---

1. Paul Lewis, "94 Nations Urge Reactor Safeguards," *New York Times,* 27 September 1986, 36.

international regime for nuclear accidents might look like and how it would operate.

THE NUCLEAR-ACCIDENT PROBLEM

Why do states find the idea of a negotiated regime for nuclear accidents attractive?[2] Is there an identifiable contract zone or zone of agreement in this realm? Are there reasons to believe that those negotiating such a regime can reach agreement on a specific outcome within such a zone? To answer these questions we must examine the characteristics of the nuclear-accident problem more closely and ask how they will influence the incentives of those considering the formation of a nuclear-accident regime.

The nuclear-accident problem is a transboundary environmental issue par excellence. The impacts of nuclear accidents are exceptionally widespread or pervasive and long-lasting. They are not limited in space or time as are the impacts of earthquakes, floods, or even famines. Unlike the effects of the 1984 Bhopal disaster, caused by the escape of a cloud of methyl isocyanate from a pesticide plant, the impacts of radiation from a severe nuclear accident are likely to be felt over hundreds or even thousands of miles.[3] In contrast to the impacts of the 1986 Rhine River disaster, caused by the flushing of toxic chemicals into the river in connection with a fire at a storage warehouse, the effects of a severe nuclear accident are apt to be felt for decades rather than years.[4] The nuclear-accident problem raises issues that clearly call for the formation of a relatively complex, global regime rather than a simple system of humanitarian disaster relief.

Additionally, the injuries resulting from nuclear accidents are demonstrable and frequently severe. It is relatively easy to detect major nuclear accidents and to track the trajectory of radioactive fallout resulting from such accidents. Though projections of the exact number of extra cancers that Chernobyl will eventually cause are debatable, the fact that the accident has already produced serious transboundary injuries is unde-

2. For a collection of essays reflecting recent thinking on international regimes more generally consult Stephen D. Krasner, ed., *International Regimes* (Ithaca: Cornell University Press, 1983).

3. On the Bhopal disaster see Sheila Jasanoff, "Managing India's Environment: New Opportunities, New Perspectives," *Environment* 28 (October 1986), 12–16 and 31–38.

4. On the Rhine River disaster see Paul Lewis, "Huge Chemical Spill in the Rhine Creates Havoc in Four Countries," *New York Times*, 11 November 1986, A1 and A6.

niable.[5] Nor are these injuries limited to predictions of possible future developments. Whereas ozone depletion or the buildup of carbon dioxide may cause severe damage in the future, the reindeer in northern Sweden affected by the radioactive fallout from Chernobyl are contaminated today.[6] In short, the nuclear-accident problem involves a clear and present danger that can be grasped without specialized training.

At the same time, the nuclear-accident problem lends itself to thinking in contractarian terms. Though every party knows that serious nuclear accidents will occur in the future, no one knows exactly when or where or how frequently they will occur. What is more, a variety of factors influence the spread of radioactive fallout from nuclear accidents so that parties cannot be certain whether they will be affected by radioactive fallout from nuclear accidents that occur elsewhere. It follows that those negotiating a nuclear-accident regime cannot know in advance what role they will occupy (for example, source, innocent victim, or fortunate bystander) in specific nuclear accidents. This means that they cannot negotiate on the basis of fixed and well-informed assumptions about their interests. Contrast this with a problem like acid precipitation in which the Scandinavians know that they are destined to be victims of sulphur dioxide generated in the British Isles and those located in eastern Canada and New England know that they are bound to be victimized by airborne pollutants originating in the American Midwest.[7] Unlike the carbon dioxide problem, the nuclear-accident problem does not even differentiate between advanced industrial states and less developed countries or between states in the northern hemisphere and those in the southern hemisphere.[8] Nuclear installations are now widely distributed in both hemispheres; many of the less developed countries find the prospect of harnessing nuclear energy for civilian purposes particularly attractive.[9] Moreover, prevailing weather and wind patterns will play a much more important role in determining who is affected by

5. For an account of some of the debates over the long-term impact of Chernobyl see Stuart Diamond, "Long-Term Chernobyl Fallout: Comparison to Bombs Altered," *New York Times*, 4 November 1984, C3.

6. Francis X. Clines, "Chernobyl Shakes Reindeer Culture of Lapps," *New York Times*, 14 September 1986, 1 and 20.

7. For an accessible account of the state of knowledge regarding acid deposition in North America see Arthur H. Johnson, "Acid Deposition: Trends, Relationships, and Effects," *Environment* 28 (May 1986), 6–11 and 34–39.

8. For an account of the current state of the debate over the greenhouse effect consult Jill Jager, "Floating New Evidence in the CO_2 Debate," *Environment* 28 (September 1986), 6–9 and 38–41.

9. Stuart Diamond, "Chernobyl Causing Big Revisions in Global Nuclear Power Policies," *New York Times*, 27 October 1986, A1 and A10.

radioactive fallout from nuclear accidents than mere geographical proximity to the site of such accidents.

These circumstances, taken together, place those negotiating a nuclear-accident regime in a situation that resembles the original position in Rawls's contractarian analysis or the precontract condition in Buchanan's discussion of constitutional contracts.[10] The parties have a good deal of general knowledge. They know, for example, that nuclear accidents will happen, that such accidents are apt to produce severe and long-lasting damages, and that these accidents will often have transboundary or even global consequences. They also know a good deal about the factors affecting the spread of radioactive fallout from nuclear accidents. But they do not know what roles they will occupy in future accidents. Specifically, they do not know whether they will be actual sites of accidents, innocent victims, or fortunate bystanders, and they do not know whether they will be in the direct path of radioactive fallout from future nuclear accidents or more peripherally located with respect to the path of the radioactive cloud.[11]

Unburdened by specific knowledge of their own roles but fully aware of the seriousness of the nuclear-accident problem, the parties are free to consider the common good as they contemplate negotiating an international regime to cope with the nuclear-accident problem. That is, they can seek to devise arrangements that would prove to be in their best interests whether they turn out to be accident sites or innocent victims, areas directly impacted or areas removed from the path of the radioactive fallout. Above all, this set of circumstances means that the parties can focus on the productive task of devising institutional arrangements to cope with the nuclear-accident problem rather than become bogged down in distributive bargaining over the locus of responsibility for specific transboundary problems or the extent to which victims are entitled to compensation from those responsible for transboundary problems.[12] Accordingly, it should come as no surprise that the Vienna negotiations conducted under the auspices of the IAEA have already produced

10. John Rawls, *A Theory of Justice* (Cambridge: Harvard University Press, 1971), and James Buchanan, *The Limits of Liberty* (Chicago: University of Chicago Press, 1975).

11. The impact of the radioactive cloud from the Chernobyl accident on Scandinavia is instructive in this context. Scandinavia probably would not have been a major impact zone at all, except for the unusual weather patterns prevailing in April and May 1986. Additionally, the impact even within Scandinavia was highly selective because of the pattern of rainfall in the area at the time.

12. On the distinction between distributive and integrative bargaining as well as other major concepts pertaining to negotiation consult Howard Raiffa, *The Art and Science of Negotiation* (Cambridge: Harvard University Press, 1982).

agreement on certain aspects of a nuclear-accident regime, whereas comparable negotiations regarding other transboundary problems are slow to make progress.

Note also that the nuclear-accident problem lends itself to calculations similar to those underlying health- or automobile-insurance schemes, a fact that suggests the establishment of a mutually beneficial international insurance regime for nuclear accidents. Of course, we can and should make a concerted effort to minimize the occurrence of nuclear accidents.[13] As with automobile accidents and major illnesses, however, we recognize that some nuclear accidents will occur, that the costs to those affected may prove overwhelming, and that the locus and timing of specific accidents are unpredictable. These are the conditions that give rise to an insurance mentality, including powerful incentives to devise arrangements to share or spread risks. Under such conditions, individual parties are willing to make limited payments on a regular basis (that is, to pay insurance premiums) to protect themselves against having to bear the full costs of a severe accident or unpredictable event. Individual participants in insurance schemes may never have occasion to file claims against their insurance. Many will not file claims equaling the amount of the insurance premiums they pay over time. But participation in various insurance schemes is worthwhile to individuals as protection against being wiped out as a result of a particularly severe accident or other unpredictable events. Once again, the contrast between this situation and other transboundary issues such as the acid-precipitation problem is striking. Whereas a group of parties may come together to create an insurance scheme to protect each member of the group against the debilitating impact of severe accidents or other unpredictable events, the acid-precipitation problem pits known victims and known sources against each other in a struggle over identifying the locus of responsibility and spelling out obligations to cover the costs of cleanup and compensation.

In the case of nuclear accidents, any insurance scheme would aim to create a general fund (that is, a kind of superfund) that all would contribute to and that would be available to cover the costs of relief and cleanup as well as the costs of compensating innocent victims. Beyond this common core, however, it is possible to envision a number of options in the formulation of an international insurance regime for

13. For an account that stresses the duty to prevent accidents see Philippe J. Sands, "The Chernobyl Accident and International Law," paper prepared for the forum on "Global Disasters and International Information Flows" organized by the Washington Program of the Annenberg Schools of Communication, 8–10 October 1986.

nuclear accidents. For instance, while a public organization, like the IAEA, might be designated or established to administer such a regime, the parties could also consider relying on private insurers operating within a framework of public regulations. Such reliance, after all, constitutes the prevailing approach to insurance schemes throughout much of the world.[14] Similarly, there are numerous possible approaches to the setting of premium schedules under such a regime. Participants might make payments based on the number of nuclear facilities they operate, the number of potential victims located nearby, their location in relationship to prevailing wind patterns, or some combination of such considerations. With the passage of time, the issue of the accident records of participants might also become a factor in the setting of premiums. A system offering lower premiums to those with good safety records would give participants an incentive to make an effort to minimize the probability of nuclear accidents occurring within their borders. A "no fault" system in which premiums were not linked to prior safety records, on the other hand, might be more acceptable to those negotiating a nuclear-accident regime in the first place on the grounds that it would minimize opportunities for invidious comparisons over time.[15]

The preceding discussion suggests that there is considerable scope for the formation of a mutually beneficial international regime designed to cope with the nuclear-accident problem. But it does not guarantee that agreement on the establishment of such a regime will be forthcoming. Situations in which identifiable contract zones exist but in which the parties nevertheless fail to reach consensus on the terms of a specific agreement occur all the time.[16] Tactical stalemates or delays attributable to hopes that the allocation of bargaining strength will shift with the passage of time often block agreement. In multilateral negotiations, like those required to cope with the nuclear-accident problem, the complexities of coalition formation and reformation add further complications to the search for agreement on specific points within contract zones or zones of agreement.

Here, too, the nuclear-accident problem exhibits characteristics that make it attractive as a focus for international regime formation. Since the beginning, there has been a tendency to treat nuclear energy as a

14. This proposition is true at the international level as well as the municipal level as the cases of shipping and air transport demonstrate.

15. The prospect of an acrimonious debate between representatives of the United States and the Soviet Union on the relative merits of the nuclear safety procedures in their respective countries makes this clear. See also Diamond, "Chernobyl Causing Big Revisions."

16. For a case study involving international environmental issues see Oran R. Young, "'Arctic Waters': The Politics of Regime Formation," *Ocean Development and International Law* 18 (1987), 101–114.

thing apart. Whether this tendency is attributable to the destructive potential of nuclear weapons, the lack of any real defense against nuclear attack, the problems posed by radioactive wastes, or some combination of these factors is difficult to determine and not of critical importance in this context. The fact is that people of all political or ideological persuasions regard nuclear energy as a unique technology that should be differentiated from other technologies and handled with greater care and restraint than other technologies.

What is more, the 1986 Chernobyl disaster has produced a distinct air of crisis regarding the nuclear-accident problem. This air is not predominantly a matter of blaming the Soviets for Chernobyl. After all, it is hard to deny (especially in view of the occurrence of other accidents such as Three Mile Island in 1979) that serious nuclear accidents could happen almost anywhere.[17] Rather, it is a matter of galvanizing world opinion behind the proposition that something must be done to improve our ability to cope with nuclear accidents. The contrast between the nuclear-accident problem and other transboundary problems is instructive in this regard also. Though it is difficult to measure such things, the carbon dioxide problem or the ozone-depletion problem could easily prove more disruptive to life on the planet over the next fifty to one hundred years than the nuclear-accident problem. Yet it is much harder to project an air of certifiable crisis surrounding the ozone-depletion problem or the carbon dioxide problem than it is with regard to the nuclear-accident problem.[18] These problems simply do not have the prominence or salience in the public imagination that problems associated with nuclear energy have. And it is hard to imagine an incident that would dramatize the sense of crisis linked to the ozone-depletion problem or the carbon dioxide problem in the same way that Chernobyl has dramatized the sense of crisis surrounding the nuclear-accident problem.[19]

These conditions go far toward explaining the momentum currently behind the nuclear-accident problem and the speed with which a large number of states have moved to negotiate at least some of the components of an international regime to cope with nuclear accidents. Not only does the atmosphere of crisis surrounding the nuclear-accident

17. See also Diamond, "Chernobyl Causing Big Revisions."

18. The recent documentation of a massive, seasonal ozone hole over Antarctica has certainly helped to draw public attention to the ozone-depletion problem and to provide the impetus behind the international agreement, signed in Montreal in September 1987, to initiate substantial cuts in the production of chlorofluorocarbons.

19. Stephen Schneider of the National Center for Atmospheric Research argues that we are now experiencing a creeping crisis in connection with the greenhouse effect. But the problem of climate change simply does not have the capacity to stir public concern that the nuclear-accident problem has in the wake of Chernobyl.

problem increase the willingness of the parties to give serious consideration to major departures from past practices, it also gives these parties clear-cut incentives to demonstrate that they are making tangible progress toward solving a pressing problem.[20] In the rush to sign agreements that prove their responsiveness to a pressing problem, however, the parties may well find themselves putting in place institutional arrangements that are less than ideal or even seriously flawed. Under the circumstances, it is important to turn at this stage to a discussion of what a fully developed international regime to cope with nuclear accidents would entail.

A REGIME FOR NUCLEAR ACCIDENTS

What do we want from an international regime for nuclear accidents? In terms of what goals or objectives should we judge recently negotiated arrangements as well as proposals for additional elements of this regime? A little reflection suggests that a nuclear-accident regime should:

- minimize the frequency with which accidents occur.
- limit damages resulting from nuclear accidents by providing for early warning, damage-control capabilities, emergency relief, and speedy clean-up procedures.
- facilitate the provision of sophisticated care for the victims of nuclear accidents.
- spread the risks associated with nuclear accidents so that individual members of international society need not bear the full costs of specific nuclear accidents.
- include arrangements for reconstruction in the wake of nuclear accidents.
- establish mechanisms for handling claims for compensation brought on behalf of those injured by nuclear accidents.
- avoid reducing the incentives of individual members of international society to make vigorous efforts to regulate the use of nuclear energy within their own jurisdictions.

With these criteria in mind, we can proceed to an examination of the major components of an international regime designed to cope with nuclear accidents.

20. Note, however, that the current atmosphere of crisis surrounding the nuclear-accident problem may have a rather short half-life. Students of public policy are well aware not only that the character of the political environment is a critical determinant of policymaking but also that political environments typically shift rapidly.

Safety Standards

Though we know that nuclear accidents will happen, it is still desirable to make every effort to minimize the probability of their occurrence. To this end it would help to articulate a common regulatory regime prescribing safety standards for the construction and operation of nuclear facilities. With respect to hardware, such standards should specify minimum requirements covering both the design and the construction of nuclear plants. With respect to operating standards, the requirements should encompass the training of personnel, operating procedures within nuclear plants, circumstances requiring temporary shutdowns, evacuation procedures, and arrangements for decommissioning. Note, however, that an international regulatory regime prescribing safety standards need not include all the features of the licensing system presently in place in the United States. The American licensing system gives rise to nuclear politics by providing what amounts to a veto to various players who have stakes in the construction and operation of nuclear facilities. The recent confrontations in a number of states over the approval of emergency evacuation plans illustrate the problems that can arise under a licensing system of this type.[21] By contrast, an international regulatory regime could prescribe clear-cut safety standards without creating opportunities for interested parties to veto the construction and operation of nuclear facilities that conform to reasonable standards.

Undoubtedly, the major problem with safety standards is to implement them in such a way as to ensure that those subject to the standards will generally comply with them.[22] Compliance can be handled in a centralized fashion through the operation of a transnational agency like the IAEA or in a decentralized fashion through the operation of responsible agencies within individual member states. Additionally, compliance can be set up either on a voluntary basis or on a mandatory basis in terms of the responsibilities of individual regime members. The current approach to this problem is a centralized but voluntary system under which the IAEA sends teams to check on compliance with safety standards only when individual states request that it do so. But this system may turn out to be inferior to a decentralized but mandatory arrangement. Such an arrangement would not pose problems during the negotiation of a regime by forcing individual states to consider transferring

21. Matthew L. Wald, "The States, the Federal Government, and the Atom," *New York Times*, 20 October 1986, A24.
22. For a general analysis of the problem of compliance as well as techniques for eliciting compliance see Oran R. Young, *Compliance and Public Authority: A Theory with International Applications* (Baltimore: Johns Hopkins University Press, 1979).

effective authority to an international agency. What is more, individual states might well prove reasonably conscientious in enforcing safety standards pertaining to nuclear facilities within their own borders. There is no real free-rider problem in this context because residents of each state would be the initial victims of laxity on the part of individual states in ensuring compliance with nuclear safety standards.[23] Furthermore, it would be perfectly possible to adjust the contributions required from individual states to reconstruction and compensation funds (more on these later) to reflect their safety records. Such a procedure would give individual states real incentives to ensure compliance with safety standards on the part of those operating nuclear facilities within their jurisdictions.[24]

Early-Warning and Damage-Limitation Procedures

Once a nuclear accident occurs, the problems of early warning and damage control arise. There is, in fact, no way to protect either natural environments or human beings from all the impacts of radioactive fallout. Yet early warning might well prove significant in minimizing the damages resulting from nuclear accidents, especially when it is coupled with well-thought-out procedures designed to protect human communities and natural environments from the effects of radiation. Precise knowledge of the scope and trajectory of the radioactive cloud would make it possible to determine what areas and populations would be placed in jeopardy directly as a consequence of any given nuclear accident. Such knowledge would also alert the public to various indirect dangers, such as the contamination of migratory birds and animals, the impact of radioactive precipitation on crops, and the passage of radioactivity through the food chain. Additionally, timely information would allow officials to consider initiating protective measures such as the temporary evacuation of human populations or the promulgation of directives regarding actually or potentially contaminated food products or other materials. Naturally, it would be desirable to set up early-warning mechanisms capable of identifying nuclear accidents immediately so as to maximize the warning time available to potential victims.

23. For a classic statement of the free-rider problem see Mancur Olson, Jr., *The Logic of Collective Action* (Cambridge: Harvard University Press, 1965). For a discussion of more recent research on this problem see Norman Frohlich and Joe A. Oppenheimer, *Modern Political Economy* (Englewood Cliffs: Prentice-Hall, 1978).

24. For a differing perspective on the issues discussed in this paragraph see Thad Cochran, "Nuclear Safety: Grounds for Optimism," *Christian Science Monitor*, 1 October 1986, 13.

But note that the problem of early warning in the case of nuclear accidents differs from the case of nuclear attack where almost all the emphasis is placed on providing warning prior to the detonation of nuclear warheads. With regard to accidents, the warning would almost always follow the release of radioactive materials. What is more, early warning might continue to prove useful over a period of days or weeks in order to provide advance notice to those located in the path of a radioactive cloud.

Early warning in connection with nuclear accidents poses an interesting technical problem that might well foster collaborative efforts among states. This concern differs fundamentally from the problem of providing early warning of nuclear attacks where individual states experience powerful competitive incentives to maximize their own early-warning capabilities and to deny such capabilities to others. With regard to nuclear accidents, in fact, it might well make sense to establish and operate joint early-warning systems.[25] This would make it possible to set up a highly dispersed array of manned and unmanned detection devices coupled with a central data-analysis facility that could piece together the larger picture from evidence derived from many sources. It would also allow for the pooling of scientific and engineering expertise in such a way as to maximize our understanding of the behavior and consequences of radioactive fallout. A positive by-product of an arrangement of this type would be the development of an international network of scientists and engineers who could quickly compare notes in the event of a nuclear accident and who would be able to facilitate effective communication among the states actually or potentially affected by an accident. The idea of establishing jointly operated early-warning systems may seem farfetched to some observers, given the political antagonisms dividing the major nuclear powers. But there is ample evidence that attitudes are now changing rapidly in this realm. The recent agreement between organizations in the United States and the Soviet Union on the exchange of on-site inspectors to monitor underground nuclear tests, for example, would have seemed out of the question even a few years ago.[26]

25. Advocates of certain types of arms-control schemes sometimes advance similar proposals. See, for example, Owen Wilkes, "A Proposal for a Demilitarized Zone in the Arctic," Project Ploughshares Working Paper 84–4.

26. The parties to the agreement are the Natural Resources Defense Council (NRDC) and the Institute of Physics of the Earth of the Soviet Academy of Sciences. See Charles B. Archambeau, *Report on the NRDC Seismic Monitoring Advisory Committee (SMAC) Meeting on May 17, 1987* (New York: NRDC, 1987).

Monitoring Facilities

Because radioactive materials are unusually long-lasting, efforts to monitor and track the impact of nuclear accidents must continue far beyond the phase of early warning. Not only is there a need to continue monitoring radioactive fallout from nuclear accidents as the resultant radioactive cloud disperses around the globe, there is also the problem of tracking the consequences of nuclear fallout as radioactive particles find their way into the food chain and eventually into the food consumed by human beings. That this can happen relatively quickly is apparent from the evidence of severe contamination in reindeer collected in northern Sweden in the aftermath of the Chernobyl disaster.[27] So far, much less is known about the long-term and cumulative impacts of radioactive fallout. But there is widespread understanding that these impacts might well pose the most serious problems associated with nuclear accidents. Coupled with evidence suggesting that the Chernobyl disaster may have emitted upwards of a quarter as much long-term radiation into the world's air, topsoil, and water as all the nuclear test explosions during the 1950s and 1960s, this aspect of the nuclear-accident problem must figure prominently in any consideration of international regime formation.[28]

The creation of suitable monitoring facilities will undoubtedly be one of the more complex tasks involved in the establishment of an international regime to cope with nuclear accidents. Such facilities would require both a network of observation posts and freedom of access for authorized personnel to all parts of the participating states. They would also require a long-term commitment to funding (immune from the fluctuations of national budgets and the political manipulations surrounding contributions to certain international organizations) as well as a central research laboratory to assimilate and interpret data collected in the aftermath of each nuclear accident. Once again, such requirements may seem unrealistic given the political atmosphere surrounding East-West relations. But there are interesting precedents for arrangements of this sort. The United States and the Soviet Union have collaborated effectively for some time on scientific research under the provisions of the Environmental Cooperation Agreement of 1972.[29] Representatives

27. Clines, "Chernobyl Shakes Reindeer Culture."
28. There has been some controversy concerning the amount of radiation emitted as a result of the Chernobyl disaster. See Stuart Diamond, "Reactor Fallout Is Said to Match Past World Total," *New York Times*, 23 September 1986, A1 and C6, and Diamond, "Long-Term Chernobyl Fallout."
29. Formally, this is the Agreement on Cooperation in Environmental Protection. Signed at Moscow in 1972, the agreement was renewed in 1977 and 1982 and reaffirmed, most recently, at the November 1985 summit meeting.

of a number of countries (including both the United States and the Soviet Union) are now cooperating enthusiastically in connection with the new International Union for Circumpolar Health. There is even a precedent for a jointly operated research laboratory in the form of the International Institute for Applied Systems Analysis (IIASA).[30] Though it would be a mistake to generalize from the specific circumstances of the IIASA, the experience of this organization is sufficient to demonstrate that we should not reject out of hand plans calling for central research facilities in connection with the nuclear-accident problem.

Relief and Reconstruction Arrangements

It is easy to sell the notion of disaster relief on humanitarian grounds. The very concept of disaster relief conjures up an image of rushing emergency supplies of food, clothing, and medicines to innocent victims of famines, floods, earthquakes, and so forth. The concept therefore suggests apolitical activities that are largely voluntary but certainly laudable on ethical grounds. In the case of nuclear accidents, by contrast, the problems of relief and reconstruction are apt to be relatively complex. Victims may be located over a wide area; providing assistance to some of them may be complicated by the presence of severe contamination, and many victims are apt to require medical attention of the most sophisticated sort. Cleanup operations following nuclear accidents may require massive efforts directed not only toward controlling dangers at actual accident sites but also toward decontamination in relatively remote areas. Moreover, there is the problem of reconstruction, especially in cases where the damaged nuclear facilities are of great economic importance. This problem encompasses an array of concerns, from the design of new installations to the handling of proposals for restarting surviving reactors in cases where accidents destroy only a portion of a nuclear complex.[31]

The key issues facing a nuclear-accident regime with regard to relief and reconstruction involve funding and administration. For reasons discussed earlier, there is much to be said for the creation of an international superfund to handle relief and reconstruction in connection with nuclear accidents. Given the scope of the damages likely to arise from serious nuclear accidents, such an arrangement is probably the only way to ensure adequate treatment of the consequences (especially the trans-

---

30. The IIASA was created in 1972 under an agreement signed by twelve academies of sciences (including five from Eastern Europe) in order to explore "ways in which science could promote international cooperation." See Raiffa, *Art and Science of Negotiation,* 3.
31. See "Soviet Plant to Resume," *New York Times,* 27 September 1986, 36.

boundary consequences) of such accidents. But this step would require those negotiating an international regime to reach agreement on a funding formula for such a superfund. Many formulas are possible, and experience at the domestic level suggests that hammering out agreement on the details of a specific formula may prove difficult.[32] In this connection, it seems critical to avoid becoming bogged down over the apportionment of relief and reconstruction costs associated with accidents that have already occurred (such as Chernobyl). The best prospect for reaching agreement is to focus on future accidents with respect to which the parties are unable to predict what their specific roles will be.

Beyond this, the administration of relief and reconstruction programs is an important and sensitive problem. Experience with disaster relief in connection with famines, floods, earthquakes, and so forth makes it clear that entrusting supplies and funds to local authorities often yields highly unfortunate results. Would the nuclear-accident problem be any different? Given the problems of containing radioactive contamination, it may be that relief and reconstruction following nuclear accidents would not lend themselves to political manipulation in the same manner that relief and reconstruction activities linked to more localized disasters are often manipulated. Nonetheless, it would be highly desirable to provide an international superfund to handle relief and reconstruction following nuclear accidents with an autonomous administration capable of making independent decisions on the apportionment of funds and of monitoring the use of these funds on the part of local authorities.

Compensation Mechanisms

Quite apart from relief and reconstruction, the nuclear-accident problem raises issues regarding liability and compensation for innocent victims of nuclear fallout.[33] From the point of view of international regime formation, the key problem here concerns innocent victims located beyond the borders of those countries in which nuclear accidents occur. Innocent victims residing within the jurisdiction where a nuclear accident occurs will undoubtedly pursue their claims in the municipal courts or legal arenas of their own country. Though this procedure may produce asymmetries in the nature or level of compensation ultimately

---

32. To illustrate, the renewal of the American superfund to clean up hazardous-waste sites (under the terms of the Comprehensive Environmental Response, Compensation, and Liability Act) was bogged down in Congress for several years over disagreements pertaining to the funding formula for the system.

33. For a good account of existing international practice regarding liability and compensation see Sands, "Chernobyl Accident."

received for comparable damages, there in nothing peculiar about this. Much the same could be said about the treatment of every category of torts in differing legal systems. But what of innocent victims residing outside the jurisdictions where nuclear accidents occur? It seems likely that governments would be reluctant to accept strict liability for damages to those residing outside their jurisdictions, especially if this arrangement came to be seen as an invitation to file large and shaky claims against foreign governments.[34] The fact that the Soviet government has recently assumed this sort of stance with regard to the Chernobyl disaster, an incident that has already occurred, is no cause for surprise. But even governments negotiating in something resembling the original position are apt to be reluctant to accept strict liability for all damages arising from nuclear accidents occurring outside their jurisdictions. On the other hand, it is hard to see why governments located in the jurisdictions where innocent victims reside should assume liability for damages resulting from nuclear accidents occurring elsewhere.[35]

What can be done to solve this problem? As suggested earlier, the problem seems to call for some sort of insurance scheme designed to generate a common fund to be used to compensate those victims of nuclear accidents who cannot turn to municipal courts or legal arenas. A little reflection suggests that we cannot rely on private enterprise to initiate an appropriate insurance scheme in this case. Among other things, the number of nuclear accidents is insufficient to permit the sort of actuarial calculations that insurers rely on to establish appropriate premium schedules. In this sense, nuclear accidents are more like acts of God than like commercial accidents affecting international shipping or air transport. Consequently, it would probably be necessary for participating governments to operate an insurance scheme designed to compensate innocent victims of nuclear accidents, making initial contributions to a common fund based on reasonable estimates but committing themselves to supplement these initial contributions if necessary. The administration of such an insurance scheme would also pose some interesting challenges to international cooperation. Given the fact that the effects of radioactive fallout can be indirect or long-term, the handling of specific claims under the joint insurance scheme suggested here could pose extremely complex evidentiary issues. While it would cer-

34. Alternatively, as Sands points out, they may react by imposing extremely low caps on maximum awards to victims. See Sands, "Chernobyl Accident," especially 32–35.

35. The Swiss reaction to the 1986 Rhine River disaster is a hopeful sign regarding the willingness of states to accept liability for injuries to innocent victims located in other jurisdictions. See Paul Lewis, "Swiss Will Pay Compensation for Spill," *New York Times*, 13 November 1986, A3.

tainly be desirable to minimize the likelihood of denying compensation to those with legitimate claims, excessively liberal standards regarding proof of injury could easily undermine the financial stability, not to mention the political acceptability, of the insurance scheme. There is also the question of whether the contributions (that is, premiums) of participating states should be adjusted periodically to reflect their nuclear safety records. A "no fault" arrangement under which premiums and safety records were not linked might lower the incentives of states to exercise strict control over their nuclear facilities. Considering that the most severe damages resulting from nuclear accidents would normally occur within the state responsible for the accident, however, it is probably a mistake to place too much stress on this concern. Additionally, it seems likely that a "no fault" arrangement would be considerably easier to sct up initially and to administer over the course of time. Contentious debates over the nuclear safety records of individual states could easily go far toward undermining the effectivenss of many elements of an international regime.

ADMINISTERING A NUCLEAR-ACCIDENT REGIME

What sorts of organizations would be required to administer a fully developed international regime for nuclear accidents? Recall that organizations are material entities possessing offices, personnel, budgets, and equipment as well as legal personality. Regimes, by contrast, are social institutions. They are sets of roles linked together by clusters of rules and rights governing the behavior of those occupying the roles. Institutions vary greatly with respect to their requirements for administrative organizations.[36] Whereas some institutions, such as competitive markets, can function smoothly with little in the way of organizations, other institutions require more or less elaborate organizations. Bear in mind, also, that organizations are costly to create and to operate. These costs encompass not only monetary expenses (for example, the expenses associated with maintaining offices and hiring staff) but also nonmonetary expenses (for example, the costs to society of supporting a bureaucracy). It follows that the burden of proof should fall on those who propose the creation of new organizations in connection with international regimes. Sometimes, the case for creating new organizations will be strong. But when the case is not strong, society will be better off

36. For an extensive analysis of relationships between international institutions and international organizations see chapter 2 above.

avoiding the burden imposed by the introduction of additional organizations.[37]

To the extent that organizations are required to operate a fully developed regime for nuclear accidents, we should give careful thought to the menu of options regarding organizations. Those negotiating the terms of a nuclear-accident regime may choose to rely on centralized organizations, such as the IAEA, decentralized organizations, such as the governments of individual member states, or some combination of the two. They may establish new organizations or turn to existing organizations, such as the IIASA, that seem capable of taking on added responsibilities. Additionally, they may encourage the emergence of private organizations, such as private insurance companies, turn to a system of public organizations, or opt for arrangements, such as public corporations, located at the intersection between private and public organizations. A complex social institution, such as the fully developed nuclear-accident regime under consideration here, may well evolve over time in such a way that it acquires an elaborate mosaic of different types of organization.

The analysis set forth in the preceding section suggests that preventative measures, such as the administration of safety standards governing the operation of nuclear facilities, may best be handled through decentralized organizational arrangements. This conclusion will surely seem counterintuitive to some observers. What sense does it make to leave the application of safety standards and, especially, the policing of such standards to the governments of those states in which nuclear facilities are located? The answer is that centralized enforcement operations are usually a comparatively small factor in determining the choices of those subject to rules or regulations regarding compliance.[38] If compliance is not forthcoming on a voluntary basis most of the time, any system of rules or regulations will fail. The key factor in this connection, then, is to make sure that the governments of individual states have adequate incentives to implement nuclear safety standards within their own jurisdictions. As mentioned previously, there is no real free-rider problem to overcome in this case. There are also possibilities for structuring the incentives of the governments of individual states through such devices as adjustments in insurance-premium schedules. But above all, governments are likely to be motivated to implement safety standards in their own jurisdictions out of a desire to avoid actions that could

37. See also Todd Sandler and Jon Cauley, "The Design of Supranational Structures: An Economic Perspective," *International Studies Quarterly* 21 (1977), 251–276.
38. Young, *Compliance and Public Authority*, especially chap. 2.

injure their own longer-term interests by initiating a process in which the entire nuclear-accident regime came unraveled.

Turning to the requirements for early warning and long-term monitoring in connection with nuclear accidents, the case for centralized organizations seems much stronger. Early warning would benefit not only from a network of manned or unmanned observation posts located throughout the world but also from a centralized facility capable of analyzing and synthesizing data reported from an array of individual observation posts. Long-term monitoring is an even more complex task, which involves sophisticated research on the consequences of introducing radioactive materials into complex ecosystems. Much of this research can undoubtedly be carried out in laboratories in individual states. But it would be of great value to have a centralized research facility as well, where those working on the long-term effects of nuclear accidents could meet to synthesize their findings and to develop an effective community of experts concerned with this problem. However, there does not appear to be a strong case for creating new organizations to handle these tasks. Existing organizations, such as the IAEA and IIASA, could easily add these tasks to their current programs. The administrative activities required to handle the early-warning and long-term monitoring operations of a nuclear-accident regime seem perfectly compatible with the current activities of organizations like the IAEA and IIASA. And such an arrangement would be comparatively efficient since it would avoid the need to establish and fund new organizations.

When we come to arrangements for a superfund to handle cleanup or reconstruction following nuclear accidents and an insurance scheme to deal with compensation for the innocent victims of nuclear accidents, the case for new organizations of a centralized nature is far stronger. These tasks would require effective organizations not only to collect revenues from participating states but also to administer substantial reconstruction programs and to process complex claims. It seems apparent that decentralized organizational arrangements would not suffice to handle tasks of this nature. They would not be able to pool the necessary funds, make legitimate decisions regarding the allocation of these funds, and respond effectively to claims for compensation advanced by those residing outside the jurisdiction of states where nuclear accidents occur. At the same time, it seems equally clear that existing organizations such as the IAEA and IIASA are not well suited to handling such tasks. These organizations are fundamentally scientific and rule making in nature. They are not set up to function as line organizations, which allocate funds and administer complex programs. Any effort to add tasks of this kind to the operations of existing organizations such as the IAEA or

IIASA would probably detract from their ability to handle their current functions without providing an effective system for tackling the new tasks.

In this era of privatization, it seems relevant to ask whether private organizations might suffice to administer the superfund and, especially, the insurance scheme associated with a fully developed international regime for nuclear accidents.[39] States could certainly make their contributions or pay their insurance premiums to private or quasi-private organizations. And many observers believe that private organizations can operate more efficiently than governmental agencies. If this is so, a move to rely on private organizations in this realm might well constitute a Pareto superior step, even allowing for a normal profit or return on investment for those operating the private organizations. Yet there are also serious drawbacks to privatization in connection with a superfund or insurance scheme created under the terms of a nuclear-accident regime. As mentioned in the preceding section, it is hard to see how the necessary actuarial calculations could be carried out in the case of nuclear accidents. Some of the participating states would represent socialist societies in which private enterprise is not a standard method of doing business. Even in capitalist systems, there is widespread agreement that industries such as the insurance industry require extensive regulation, a fact that casts doubt on the alleged benefits of private arrangements in terms of efficiency, at least in this domain. Accordingly, a straightforward private-enterprise system is not a likely outcome. It might well be appropriate, however, to explore the prospects for intermediate organizations, like public corporations, in this realm. Even committed capitalist systems, such as the United States, have extensive experience with public corporations in such forms as the Federal Deposit Insurance Corporation, the Tennessee Valley Authority, and the Commercial Satellite Corporation.[40] At the international level, we might examine such models as the Bank for International Settlements and the International Satellite Corporation in the search for relevant precedents.

Conclusion

The nuclear-accident problem exhibits a number of characteristics that make it an attractive focus for regime formation at the international

39. For a collection of essays exploring the privatization movement from a number of perspectives see *Journal of Policy Analysis and Management* 6 (Summer 1987).

40. For a review of the experience with the Tennessee Valley Authority see S. David Freeman, "The Nine Lives of TVA," *Environment* 27 (April 1985), 6–11.

level. The international community has already moved to seize the resultant opportunity with the signing in September 1986 of conventions dealing with early-warning mechanisms and emergency-relief systems for nuclear accidents. Yet this initial success does not insure that the emerging regime for nuclear accidents will be optimal or even complete. There is a tendency to think of the nuclear-accident problem in terms of conventional ideas relating to disaster relief. But nuclear accidents differ from earthquakes, floods, and even famines in significant ways. They constitute transboundary environmental hazards par excellence. Their impacts are not only likely to be widespread, they are also long-lasting. It follows that the nuclear-accident problem poses requirements for cleanup or reconstruction procedures and compensation mechanisms well in advance of those arising from more conventional disasters. Additionally, the nuclear-accident problem raises interesting issues relating to the creation of new organizational arrangements, such as a superfund to handle cleanup and reconstruction or an insurance scheme to deal with claims for compensation. Though such organizations would be distinctly innovative at the international level, they are not entirely without precedent. Note also that the Chernobyl incident, fueling the pervasive public fear of nuclear technology, has stimulated a strong sense of momentum surrounding the effort to develop an effective international regime to deal in an orderly fashion with an array of issues associated with nuclear accidents. Under the circumstances, it would be unfortunate to miss an historic opportunity by settling for a rather conventional nuclear-accident regime rather than capitalizing on this momentum to take some bold steps toward the creation of innovative international institutions.

CHAPTER SEVEN

Regime Formation as Conflict Resolution: Arctic Shipping

Since the S/T *Manhattan*, an ice-strengthened American supertanker, steamed through the Northwest Passage in 1969, Canada and the United States have engaged in a running conflict over the status and use of Arctic waters.[1] Canada reacted swiftly and unilaterally to the voyage of the *Manhattan*, asserting authority to control pollution in Arctic waters by enacting the Arctic Waters Pollution Prevention Act of 1970. The United States has consistently refused to acknowledge the validity of this legislation, at least as far as non-Canadians are concerned. Canadian representatives lobbied hard at sessions of the Third United Nations Conference on the Law of the Sea through the 1970s for the inclusion of Article 234, a special provision extending the regulatory authority of coastal states in ice-covered areas. American representatives pressed equally hard for provisions guaranteeing nonsuspendable transit rights in the articles pertaining to straits used for international navigation (Articles 34–45).

The 1980s have witnessed an intensification of this conflict. For starters, the United States refused to sign the completed Law of the Sea Convention in 1982, raising doubts about the future status of this convention with regard to Canadian-American relations. Then, in the summer of 1985, the United States Coast Guard icebreaker *Polar Sea* transited the Northwest Passage en route from Thule, Greenland, to Barrow, Alaska, without seeking Canadian permission. The Canadian government promptly responded by issuing a formal declaration claiming all the waters of the Canadian Arctic Archipelago as internal waters of

---

1. For relevant background see Franklyn Griffiths, ed., *The Politics of the Northwest Passage* (Montreal: McGill-Queen's University Press, 1987).

Canada and announcing plans to build the *Polar 8,* a vessel that would become the most powerful icebreaker in the world. Following up on these initiatives, Canada published a defense white paper in June 1987 calling for the acquisition of a fleet of ten to twelve nuclear-powered attack submarines intended, at least in part, to provide Canada with an ability to assert its jurisdiction in the Arctic.[2]

The thesis of this chapter, nonetheless, is that the Canadian-American conflict over Arctic shipping is more apparent than real. This putative conflict is largely an artifact of the propensity of the parties to formulate issues relating to Arctic shipping in jurisdictional terms, thereby forcing themselves to address the problem in the mode of distributive bargaining. Yet Canada and the United States need not adopt this approach. There is nothing to prevent the two countries from setting jurisdictional questions aside without prejudice and proceeding to develop an international regime for Arctic shipping. Approaching the matter in this way would open up numerous opportunities to explore mutual interests through a process of integrative bargaining in contrast to distributive bargaining.[3]

PROSPECTS FOR ARCTIC SHIPPING

Swept along by evidence suggesting the technical feasibility of Arctic shipping and by the expansive spirit underlying the idea of a "great Arctic energy rush," many observers simply assume that the next two or three decades will witness a rapid growth in Arctic shipping.[4] But this is hardly a foregone conclusion. Even among those who confidently expect marked growth in Arctic shipping, moreover, the relative merits of ice-breaking supertankers and submarine tankers of varying proportions are far from settled.

The future of Arctic shipping rests on a series of factors that will determine the course of nonrenewable-resource development (and es-

2. At this writing (December 1987), the two countries are moving toward an agreement on the use of the Northwest Passage by American icebreakers, which would accommodate some Canadian concerns about jurisdiction without requiring a formal acknowledgement of Canadian sovereignty over Arctic waters on the part of the United States.

3. On the distinction between distributive bargaining and integrative bargaining see Howard Raiffa, *The Art and Science of Negotiation* (Cambridge: Harvard University Press, 1982), pts. 2 and 3.

4. "Great Arctic Energy Rush," *Business Week,* 24 January 1983, 52–57. See also T. C. Pullen, "Arctic Outlet," an address to the Canada-Japan Trade Council of Ottawa, Ottawa, 1981.

pecially oil and gas development) in the region. These include the magnitude of recoverable reserves of hydrocarbons in the Arctic, the attractiveness of Arctic oil and gas in economic or commercial terms, the political attractions of the nonrenewable resources of the Arctic, and the relative merits of alternative methods of transporting Arctic resources to southern markets. Significant uncertainties regarding each of these factors make it difficult to predict the future of Arctic shipping with any confidence.

Many commentators believe that up to 50 percent of the earth's remaining undiscovered reserves of hydrocarbons are located north of 60°N latitude. Since the massive strike at Prudhoe Bay in 1968, however, discoveries of proven reserves of oil and gas in the North American Arctic have been few and far between. Several significant (though much smaller) fields have turned up in areas adjacent to Prudhoe Bay, and scattered deposits of gas and oil have been located in the Canadian high Arctic islands and Beaufort Sea.[5] But there have also been major disappointments in areas widely expected to contain sizable deposits of oil and gas (for example, the U.S. lease sale number 71 area). At the same time, exploratory efforts in several areas outside the Arctic (for example, eastern Canada, the Gulf of Mexico, the South and East China Seas) have produced promising evidence of large recoverable reserves of hydrocarbons. None of these facts demonstrates that there will be no Arctic energy rush in the foreseeable future. But they do remind us of the profound uncertainties surrounding the search for hydrocarbons in areas like the Arctic.

No matter how large the recoverable reserves of the Far North, Arctic oil and gas will always be expensive to produce.[6] This is a function of the severity of Arctic operating conditions together with the problems of transporting Arctic hydrocarbons to distant markets, and these cost factors are not subject to change. As a result, the commercial attractiveness of Arctic oil and gas will be particularly sensitive to fluctuations in supply and demand in world markets. If demand fails to rise steadily (owing to such things as energy conservation) or if supplies prove plentiful (owing to the development of alternative energy sources or discoveries of oil and gas elsewhere), the marginal hydrocarbons of the Arctic

5. John C. McCaslin, ed., *International Petroleum Encyclopedia* (Tulsa: Pennwell Publications, 1984); Bob Williams et al., "North American Arctic Report," *Oil and Gas Journal*, 25 June 1984, 55–77; David B. Brooks, "Black Gold Redrilled: Are the Economics of Beaufort Sea Oil Getting Better or Worse?" *Northern Perspectives* 11, no. 3 (1984), 1–4; and M. B. Todd, "Development of Beaufort Sea Hydrocarbons," *The Musk-Ox* 32 (1984), 22–43.

6. For relevant background consult J. W. Devanney III, *The OCS Petroleum Pie*, MIT Sea Grant Program, Report No. MITSG 75–10 (Cambridge, 1975).

will be among the first to become unattractive in purely commercial terms. What is more, the fact that Arctic oil and gas is expensive to produce and ship makes Arctic development economically vulnerable to the activities of organizations like OPEC, which influence world market prices for oil and gas over and above the normal interaction of supply and demand. Because most OPEC oil is relatively cheap to produce and deliver, OPEC could destroy the commercial attractiveness of Arctic oil and gas should it choose to do so. It is therefore hard to escape the conclusion that the development of Arctic oil and gas reserves will always be risky in purely commercial terms.

What ultimately drives the search for Arctic oil and gas is the attractiveness of this energy source in political terms. Because the North American Arctic is controlled by the United States and Canada, many Americans regard the development of Arctic hydrocarbons as a promising route to energy independence.[7] But even this impetus does not ensure that we will witness an Arctic energy rush during the foreseeable future. Many thoughtful Americans are more interested in the pursuit of energy conservation, enhanced efficiency, and energy alternatives than in deploying ever more advanced technology to conquer the Arctic. The development of the Strategic Petroleum Reserve has ameliorated the fears of many concerning new oil embargoes.[8] Both environmentalists and Native peoples are presenting powerful arguments in important political arenas concerning the disruptive potential of oil and gas development under Arctic conditions. For its part, Canada has ample energy reserves outside the Arctic to ensure self-sufficiency for itself for years to come.[9] Given the influence of nationalist sentiment in Canada in recent years, moreover, we cannot simply assume that Canadian governments will be moved to join an Arctic energy rush in order to promote energy independence for the United States. Therefore, though the political attractions of Arctic oil and gas are real, they are not so overwhelming that we can confidently predict that recoverable reserves of oil and gas in the Arctic will be developed at a rapid pace.

Additionally, the relative merits of alternative methods of transporting Arctic oil and gas are anything but settled.[10] Many analysts suggest

7. John Dyson, *The Hot Arctic* (Boston: Little, Brown, 1979).
8. McCaslin, *International Petroleum Encyclopedia*, 316–322.
9. John F. Helliwell, Mary E. MacGregor, and Andre Plourde, "Changes in Canadian Energy Demand, Supply, and Politics," *Natural Resources Journal* 24 (1984), 297–324.
10. Consult, inter alia, O. S. C. Robertson, "The Canadian Arctic Water Basin" and "The Russian Arctic Water Basin," both in Maja van Steensel, ed., *People of Light and Dark* (Ottawa: Queen's Printer, 1966); T. C. Pullen, "The Development of Arctic Ships," in Morris Zaslow, ed., *A Century of Canada's Arctic Islands* (Ottawa: Royal Society of Canada, 1981); and A. S. McLaren, "Transporting Arctic Petroleum: A Role for Commercial Submarines," *Polar Record* 22 (1984), 7–23.

that marine transport, whether by ice-breaking supertanker or by submarine tanker, has great attractions in terms of flexibility. It is possible for ships to roam widely, loading oil or gas even from relatively small fields. At the same time, it appears that pipelines are more reliable than marine transport and may continue to be for years to come. Perhaps because pipelines have become a fact of life in the North, the opposition of environmentalists and Native peoples to proposals for new pipelines is now somewhat less vociferous than opposition to plans for marine transport.[11] What is more, there are many cases in which it is possible to configure pipelines in such a way that they lie wholly within the jurisdiction of a single state, thereby avoiding the inevitable complications of international coordination of the transport of oil and gas.

Within the framework of this broad picture, there are significant national variations regarding the prospects for Arctic shipping. The political attractions of Arctic oil and gas are undoubtedly greater for the United States than for Canada. The United States is a large energy importer possessing few new sources of recoverable reserves of hydrocarbons outside the Arctic. Canada, by contrast, is largely energy self-sufficient, and it has excellent prospects for developing new reserves of hydrocarbons outside the Arctic (for example, the Hibernia area off Newfoundland or the Sable Island area off Nova Scotia).[12] On the other hand, marine transport is apt to hold greater attractions for the movement of oil and gas from the Canadian Arctic than from the American Arctic. Reserves in the Canadian Beaufort Sea and the high Arctic islands are likely to take the form of relatively small deposits scattered over an enormous area. American Arctic reserves are more likely to consist of a few large deposits located in the Beaufort fringe or the coastal plain of the Arctic National Wildlife Refuge, where they can be tied into existing or proposed pipeline systems focused on Prudhoe Bay. None of these considerations proves that those who foresee a substantial growth of Arctic shipping in the next two or three decades will turn out to be wrong. But it certainly does license the observation that developments along these lines hardly constitute a foregone conclusion.

SOURCES OF CONFLICT

Nonetheless, both Canada and the United States have vital interests that could be affected by any significant development of Arctic ship-

---

11. Francois Bregha, "CARC's Memorandum to Cabinet," *Northern Perspectives* 10, no. 3 (1982), 1–12.
12. Williams et al., "North American Arctic Report."

ping.[13] Several of these interests will be operative under virtually any foreseeable circumstances. Others will vary in their import as a function of such things as the course of East-West relations, the tone of North-South interactions, and the tides of nationalism or protectionist sentiment in Canada and the United States. It is the configuration of these interests that has led many observers to anticipate an increasingly severe conflict between Canada and the United States over Arctic shipping.

Political Interests

For shorthand purposes, we can use the term *sovereignty* to characterize Canada's political interests in the high Arctic.[14] Sovereignty is partly a matter of natural sensibilities regarding the physical integrity of the nation.[15] Under the most optimistic of assumptions, Canada's presence in the high Arctic will remain thin for the indefinite future. It is therefore not totally unrealistic to imagine threats to Canada's "effective occupancy" of the region. In part, the concern for Arctic sovereignty is a deep-seated symbolic one. Because many Canadians think of Canada as a northern country, it is important to their sense of national identity to exert the widest possible control over the high Arctic. For this reason, commentators frequently treat northern sovereignty as a core objective of Canadian foreign policy.

In fact, no one questions Canadian sovereignty over the islands in the Canadian sector of the Arctic.[16] The critical issue, therefore, concerns control over the waters of the Canadian Arctic Archipelago (some would extend this to the waters of the Arctic sector stretching northward to the Pole). Many Canadians have favored the assertion of total control over these waters, claiming them as internal waters of Canada or at least as parts of the Canadian territorial sea. In 1985, in fact, the Canadian government acted formally to enclose the waters of the Arctic Archipelago as internal waters. Yet Canada has other political interests that cut against such an expansive assertion of control. Should Canada persist in this course, others (such as the United States, Denmark/Greenland, and the Soviet Union) will be encouraged to follow suit in their

13. See E. J. Dosman, ed., *The Arctic in Question* (Toronto: Oxford University Press, 1976); Lisle A. Rose, "Recent Trends in U.S. Arctic Policy," *Arctic* 35 (1982), 241–242; and R. Tucker Scully, "Arctic Policy: Opportunities and Perspectives," in *Proceedings of a Conference on Arctic Technology and Policy* (New York: Hemisphere Publishing, 1983).

14. See the essays in Dosman, *Arctic in Question*.

15. R. St. J. Macdonald, ed., *The Arctic Frontier* (Toronto: University of Toronto, 1960).

16. Dosman, *Arctic in Question;* Brian D. Smith, "United States Arctic Policy," University of Virginia Ocean Policy Study 1:1 (Charlottesville: Michie, 1978), and Hal Mills et al., *Ocean Policy and Management in the Arctic* (Ottawa: CARC, 1984).

sectors of the Arctic. Developments along these lines might well yield greater losses than gains for Canada in the long run.[17] What is more, taking the lead in the enclosure movement for marine areas runs counter to Canada's long-standing policy of promoting international cooperation in many realms. Therefore, while adopting a relatively extreme position on Arctic sovereignty may satisfy the nationalistic aspirations of certain Canadians at the present time, such a posture could prove costly in terms of Canada's larger role and reputation in the international community.

The political interests of the United States in the high Arctic, by contrast, can be captured in the terms *security* and *energy independence*.[18] Until recently, it was widely held that the strategic significance of the Arctic had waned as a consequence of the decline of the manned bomber as a delivery vehicle for strategic weapons. But several recent developments have radically altered this situation, focusing renewed interest on the Arctic in terms of security.[19] These developments include the increased vulnerability of land-based ballistic missiles, the deployment of cruise missiles on manned bombers, and the growing importance of the Arctic as a theater of operations for nuclear-powered submarines. As a result, the United States has moved vigorously to initiate such projects as the construction of a North Warning System (an upgraded version of the DEW Line of the 1950s) and the development of more sophisticated antisubmarine warfare capabilities for use in the Arctic. What all this means is that it is exceedingly unlikely that the United States will acquiesce during the foreseeable future in any form of Canadian control over the waters of the Arctic Archipelago that substantially reduces the transit rights of American vessels, allows Canada to require submarines to pass through Arctic waters on the surface,[20] or limits the use of the superjacent airspace.

No doubt, the Arctic is less vital to the United States in terms of energy independence than in terms of security. Conservation and alternative

17. The disadvantages to Canada of having the entire Arctic carved up into an elaborate network of national zones might well outweigh any advantages Canada might obtain from extending its own jurisdiction in the Arctic. On the realism of this scenario see also W. Joseph Dehner, "Creeping Jurisdiction in the Arctic: Has the Soviet Union Joined Canada?" *Harvard International Law Journal* 13 (1972), 271–288.

18. National Security Council, "National Security Decision Memorandum 144" (Washington, D.C., 1971), and William E. Westermeyer and Kurt M. Shusterich, eds., *United States Arctic Interests: The 1980s and 1990s* (New York: Springer-Verlag, 1984).

19. For a broader discussion of Arctic security issues consult Nils Orvik, *Northern Development, Northern Security* (Kingston: Queen's University, 1983).

20. Any such requirement would effectively preclude submarine operations in the Arctic.

energy sources are proving to be live options. Substantial new oil and gas reserves are being found in other parts of the world. The Strategic Petroleum Reserve is becoming a reality. Yet the goal of energy independence is a powerful political force in the United States, and the idea of the Arctic as a storehouse of hydrocarbons is firmly implanted in the public mind. It is therefore unrealistic to expect the United States to accept jurisdictional arrangements in the Arctic that could curtail the ability of American companies to move Arctic oil and gas to southern markets. Any such move would appear to undermine the very features of the Arctic that make Arctic oil and gas politically attractive.

Economic Interests

Both Canada and the United States have general economic interests as well as specific economic interests in the Arctic. The United States may come to regard Arctic oil and gas as a secure source of energy, which will provide protection against energy shocks like those of the 1970s and contribute to a continued growth of productivity in the American economy. Canada may exploit Arctic hydrocarbons to develop an export market in the hope of maintaining a vibrant Canadian economy. In more specific terms, the Canadian government may experience irresistible pressure to promote Arctic oil and gas development to prevent a total collapse of Dome Petroleum or to shore up Panarctic Oils (in which the Canadian federal government has a large stake).[21] For its part, the government of the United States may experience powerful pressures to bail out companies whose Arctic ventures have run into serious trouble. The efforts of the Reagan administration to facilitate the financing of the Alaska Natural Gas Transport System illustrate this prospect.

Is there any reason to expect these economic interests to lead to a Canadian-American conflict over Arctic shipping? Restrictive Canadian regulations for marine transport in the Arctic could certainly become an irritant. Unwillingness on the part of American financiers to provide funds for future Canadian projects involving Arctic shipping (such as the now defunct Arctic Pilot Project) could cause friction. The Canadian government could even seek to restrict American shipping in the Arctic for the express purpose of forcing the United States to import Canadian oil or gas on Canadian terms. Yet all these possibilities seem a little remote and unpersuasive. In reality, the strictly economic interests of the two countries in Arctic shipping do not appear to be incompatible.

21. See Gurston Dacks, *A Choice of Futures: Politics in the Canadian North* (Toronto: Methuen, 1981), especially chap. 4.

In contrast to the case of political interests, there is nothing in these economic interests that cannot be worked out handily through the efforts of technical experts.

Environmental Interests

The fact that large-scale shipping has never occurred in the North American Arctic makes it impossible to evaluate the probable environmental impacts of such shipping on the basis of experience. Nonetheless, there are good reasons to be highly sensitive to the potential environmental disruptions of Arctic shipping.[22] The Arctic exercises a powerful influence on global climate patterns; the characterization of the Arctic as the world's "weather kitchen" is apt. Arctic ecosystems are undoubtedly fragile in some important respects. It takes extraordinary lengths of time, for example, for Arctic ecosystems to recover from serious disruptions. Above all, many Arctic ecosystems are poorly understood so that it is impossible to make confident predictions regarding the environmental impact of Arctic shipping.

Without doubt, concern about Arctic environmental matters has risen rapidly in recent years in Canada and in the United States, and it seems fair to say that this concern is now becoming institutionalized in both political systems. It is possible to argue about the relative impact of environmental considerations in the policy-making processes of the two countries (for example, is the American EIS process more or less effective than the Canadian EARP process?). But there is no denying the fact that environmental interests have become important factors in both Canadian and American policy-making. Even more important for this discussion, however, there is no reason to anticipate any serious conflicts between Canadian and American environmental interests regarding Arctic shipping. Canada has shown great concern about the environmental integrity of the Lancaster Sound area.[23] The United States has displayed a similar sensitivity to the welfare of the bowhead whale stocks of the western Arctic.[24] On the other hand, environmental concerns may not have been accorded sufficient weight in the Canadian controversy over the construction of port facilities at Stokes Point.[25] Much the same can be said regarding the American controversy over seasonal

22. For a brief but informative account see Central Intelligence Agency, *Polar Regions Atlas* (Washington, D.C.: GPO, 1978).
23. See the essays in E. F. Roots, ed., *Lancaster Sound: Issues and Responsibilities* (Ottawa: CARC, 1980).
24. David Boeri, *People of the Ice Whale: Eskimos, White Men, and the Whale* (New York: Dutton, 1984).
25. Peter Burnet, "Stokes Point, Yukon," *Northern Perspectives* 11, no. 2 (1983),1–12.

drilling restrictions in the Beaufort Sea.[26] In short, both countries are clearly concerned about the potential environmental consequences of Arctic shipping, but there is no guarantee that environmental interests will dominate policy-making in either country.

Social Interests

During the 1970s and 1980s, the Inuit (Eskimo) residents of the high Arctic acquired an increasingly effective voice in plans for Arctic development. This trend is partly a result of growing receptivity in American, Canadian, and Danish societies to the claims of minorities phrased in terms of rights. It is no longer politically acceptable in any of these societies simply to ignore the rights of minority groups in the interests of serving the needs of dominant groups. In part, the trend stems from the growing financial and political capabilities of the indigenous peoples of the North American Arctic. Both the financial resources of the North Slope Borough in Alaska and the significance of the Inuit Circumpolar Conference as a political forum for the articulation of Native interests are beyond doubt.[27]

What this development means in the present context is that the potential impact of Arctic shipping on the remote communities of the Arctic cannot be ignored. If anything, the Inuit across the Arctic are more vocal and united in their opposition to Arctic shipping than they are to Arctic pipelines. A single shipping route could produce significant impacts on communities from Greenland to Alaska; pipelines seldom affect such a broad array of peoples. What is more, it is reasonable to expect that Arctic shipping would have more severe impacts on subsistence practices of indigenous peoples than would pipelines. Most of these peoples have developed maritime adaptations; they are "hunters of the northern ice."[28] Additionally, existing pipelines have compiled a relatively good record in these terms; Arctic shipping may well generate unpredictable but potentially frightening consequences. What this means is that both the Canadian government and the government of the United States will experience strong pressure to weigh social consequences heavily in their

26. "State Lifts Beaufort Broken-Ice Restrictions," *Arctic Policy Review* 2, no. 6 (1984), 9–13.

27. Gerald MacBeath, *North Slope Borough Government and Policymaking*, MAP Monograph No. 3 (Anchorage: Institute of Social and Economic Research, 1981), and Philip Lauritzen, *Oil and Amulets* (St. Johns: Breakwater Books, 1983).

28. Richard Nelson, *Hunters of the Northern Ice* (Chicago: University of Chicago Press, 1969), and Alan Cooke and Edie Van Alstine, eds., *Sikumiut: "The People Who Use the Sea Ice"* (Ottawa: CARC, 1984).

decisions on Arctic shipping. But this fact does not suggest the emergence of serious conflict between the two countries over Arctic shipping. On the contrary, the United States and Canada will experience significant impacts to cooperate in an effort to minimize social impacts from any large-scale Arctic shipping.

Scientific Interests

It is probably fair to conclude that scientific research in the North American Arctic has been less extensive over the last several decades than research in Antarctica.[29] But this situation may soon begin to change. The Arctic Research and Policy Act of 1984 (PL 98–373) should give a boost to Arctic research in the United States. Similar pressures to increase support for Arctic research in Canada may produce results in the near future.[30] Of course, the expansion of research may contribute to the intensification of conflict if research programs are harnessed to nationalistic goals. It is a cause for some concern that recent Canadian work in the field of marine geology (for example, the exploration of the Alpha Ridge) is closely linked to Canadian interests in expansive assertions of northern sovereignty.[31] Similarly, the fact that American Arctic research has often been tied closely to the security interests of the United States is hardly a source of comfort.

Yet the rise of serious interest in Arctic shipping could provide a great impetus for Arctic research in a number of fields and serve as a basis for fruitful Canadian-American collaboration in the Arctic. Whatever our preferences may be, most research is stimulated by practical concerns. Undoubtedly, the growth of shipping over the Northern Sea Route has been a major factor underlying the importance attached to Arctic research in the Soviet Union.[32] Regular shipping in the North American Arctic would produce a pressing need for better information about such matters as hydrology, ice dynamics, weather patterns, and Arctic ecosystems. Far from being a source of conflict, the resultant research programs could serve to promote Canadian-American cooperation with respect to Arctic shipping. The research needs of authorities in both

29. David M. Hickok et al., "United States Arctic Science Policy," Anchorage: Alaska Division of the American Association for the Advancement of Science, 1981.
30. See W. P. Adams, P. F. Burnet, M. R. Gordon, and E. F. Roots, *Canada and Polar Science* (Ottawa: Department of Indian Affairs and Northern Development, 1987).
31. Ken MacQueen, "Canada Probes Arctic Ice to Back Claims to Ownership," repr. in *Polar Times* 96 (June 1983), 11.
32. William E. Butler, *Northeast Arctic Passage* (Alphen aan den Rijn: Sijthoff and Noordhoff, 1978).

countries would be much the same. It would make sense to avoid unnecessary duplication of expensive Arctic research. Scientists also have an outstanding record of being able to collaborate productively, even in cases where their national governments are at odds.

This brief account of the prospects for Arctic shipping over the next several decades as well as the character of Canadian and American interests in Arctic shipping does not warrant the conclusion that this issue area is likely to precipitate a severe crisis in Canadian-American relations in the foreseeable future. Even so, there is a persuasive case for coming to terms now with issues pertaining to Arctic shipping. Though Arctic shipping has the potential to generate relatively severe frictions between the two countries, no genuinely intractable conflicts have arisen in this issue area to date. The fact that large-scale Arctic shipping is not an immediate prospect makes it possible to consider the issues with a certain calmness and objectivity. As I shall show in the remaining sections of this chapter, there are solutions to the outstanding issues that would secure the vital interests of both countries and provide for management strategies to meet the needs of those actually engaged in, or affected by, Arctic shipping. No doubt, the occasionally abrasive character of Canadian-American relations in recent years could become a barrier to the successful resolution of outstanding issues relating to Arctic shipping. Approached from another angle, however, success in resolving Canadian-American differences over Arctic shipping could contribute significantly to the initiation of a constructive dialogue between the two countries on a number of other issues (for example, acid precipitation, fisheries management, defense installations, economic imperialism versus economic nationalism).[33]

THE POLITICS OF JURISDICTION

Much of the air of conflict surrounding Arctic shipping stems from the propensity of the parties to formulate issues in jurisdictional terms. Broadly, jurisdiction involves the demarcation of boundaries within which authority may be exercised.[34] Many jurisdictional boundaries are geographical. Debates concerning the breadth of the territorial sea or the locus of baselines demarcating the inner boundary of the territo-

---

33. See also Lincoln P. Bloomfield, "The Arctic: Last Unmanaged Frontier," *Foreign Affairs* 60 (1981), 87–105.

34. For a standard treatment consult J. L. Brierly, *The Law of Nations*, 6th ed. (New York: Oxford University Press, 1963).

rial sea highlight this class of boundaries. Alternatively, jurisdictional boundaries may be demarcated in functional terms. States may assert jurisdiction over specific functions, such as outer continental-shelf oil and gas development or pollution control, without claiming jurisdiction over other activities in the same area. It is conventional to speak of sovereignty or sovereign authority when a state exercises authority over the full range of functional activities occurring in a well-defined geographical area.

These observations make it clear that bundles of jurisdictional claims are not indivisible. There is nothing to prevent one state from exercising authority over certain activities (for example, commercial fishing) in a given area while other states exercise authority over other activities (for example, maritime transport) in the same area. By way of illustration, such a division is precisely the situation that has arisen in recent years with regard to fishery-conservation zones and even exclusive economic zones. When the relevant activities interfere with one another, of course, it is necessary for the states involved to work out some system of rights and rules to resolve the resultant conflicts. Whether they take the form of informal understandings or explicit agreements articulated in formal treaties, the operation of these systems of rights and rules gives rise to international regimes.

There are several reasons why the formulation of issues relating to Arctic shipping in jurisdictional terms has the effect of heightening Canadian-American conflict and promoting undesirable outcomes. At best, the resolution of jurisdictional conflicts is apt to become a matter of distributive bargaining. Worse yet, jurisdictional differences may well give rise to disputes in which there is no contract zone or zone of agreement in the positions of the parties.[35] Consider the Canadian-American dispute over Arctic shipping in this light. Canada insists that the waters of the Northwest Passage and the Arctic Archipelago more generally should be treated as internal waters of Canada. But nonsuspendable transit rights and rights to use the superjacent airspace are not guaranteed in internal waters. Given the interests of the United States in the high Arctic, it follows that the United States cannot accede to Canada's claim regarding internal waters. Any disposition of the problem formulated in this fashion would therefore require one party to become a clear-cut loser. Not only is it unlikely that either party will accept such an outcome voluntarily, the emergence of such an outcome is also likely to prove costly in terms of the future of Canadian-American relations more broadly.

35. For a discussion of these concepts consult Raiffa, *Art and Science of Negotiation*, pt. 2.

Even where there is a zone of agreement regarding jurisdictional issues, the resultant bargaining will generally take the form of distributive politics. In its essentials, that it, the situation will resemble a negotiation between a buyer and a seller over the purchase price of a house or between labor and management over wage rates. There may be certain outcomes that both sides prefer to an outcome of no agreement. But each party will concentrate on devising tactics designed to procure the best possible outcome for itself in distributive terms rather than developing new opportunities for mutually beneficial relationships. Any jurisdictional claims that the United States concedes to Canada would constitute a deterioration in the "terms of trade" from an American point of view and vice versa. As is well-known, distributive bargaining of this type routinely produces a preoccupation with bargaining tactics and often leads to protracted stalemates, which are costly to all parties concerned.[36] Because such interactions are frequently perceived as tests of bargaining strength, moreover, participants regularly choose to sacrifice the prospects of gains in specific situations in order to avoid any appearance of weakness, which might prompt others to press harder for concessions in concurrent or subsequent negotiations.

Even if the United States and Canada were to reach some agreement regarding the jurisdictional status of the waters of the Northwest Passage and the Arctic Archipelago, this would do nothing to solve many of the major management problems that would arise in connection with regular Arctic shipping. Here there are two distinct cases to consider: the case of ice-breaking supertankers and the case of submarine tankers. Most plausible routings would take ice-breaking supertankers (carrying either oil or liquid natural gas) through the jurisdictional zones of two or more of the following states: Denmark/Greenland, Canada, and the United States. It follows that none of the North American Arctic states can use this mode of transport without explicitly coordinating its management practices with one or both of the others. What this suggests is the importance of joint problem solving with regard to Arctic shipping, in contrast to the strategic maneuvering characteristic of distributive bargaining. This point has particular significance for Canadian policymakers because the most probable impetus to the introduction of ice-breaking supertankers in the Arctic during the foreseeable future will be efforts to move Canadian hydrocarbons to the east coast through the jurisdictional zone of Denmark/Greenland or to Japan through the jurisdictional zone of the United States.

36. Thomas C. Schelling, *The Strategy of Conflict* (Cambridge: Harvard University Press, 1960).

The case of submarine tankers raises different management problems because these tankers would not be confined to the waterways of the Northwest Passage, restricted areas of Baffin Bay, and the Beaufort Sea fringe.[37] Submarine tankers taking on oil or gas at terminals in Alaska, Canada, or Greenland might well be able to deliver their cargoes to southern markets without entering the jurisdictional zones of other countries. A partial exception to this might occur if every Arctic state were to advance and enforce the most extreme sort of sector claim.[38] Even this might complicate the situation more for submarine tanker routes originating in Canada than for routes originating in Alaska or Greenland. What is important to notice about this case, however, is the fact that the dangers of marine pollution would not decline, despite the ability of shippers to escape the regulatory requirements of one or more of the North American Arctic states. This is so because of the natural conditions prevailing in the Arctic. Oil takes years or even decades to degrade under Arctic conditions, and spilled oil (especially spilled oil trapped in or under ice) would tend to circulate around the Arctic Basin just as ice islands do today.[39] What all this means is that a Canadian-American agreement regarding the jurisdictional status of the waters of the Northwest Passage and the Arctic Archipelago would do little to solve the real management problems associated with submarine tanker traffic in the Arctic. If efforts to reach such an agreement produced an atmosphere of strategic maneuvering and abrasive conflict, moreover, the prospects for dealing with the problems of submarine-tanker traffic in a cooperative mode would decline.

What is more, the resolution of outstanding issues relating to Arctic shipping through the specification of some system of jurisdictional zones would constitute a bad precedent, or even a step in the wrong direction, when it comes to handling other important Arctic issues. The dangers to Arctic ecosystems associated with maritime transport are real. But several other environmental concerns are equally important, if not more important, in the Arctic today. Arctic haze has become a severe problem, which could have far-reaching impacts on global weather patterns. Oil spills and chronic discharges from offshore platforms are at least as worrisome as the potential disruptive effects of Arctic shipping. The industrialization of the Arctic poses numerous threats to the welfare of marine-mammal stocks. A common feature of all these problems is that

37. McLaren, "Transporting Arctic Petroleum."

38. See also Donat Pharand, *The Law of the Sea of the Arctic* (Ottawa: University of Ottawa Press, 1973), and Mills et al., *Ocean Policy and Management.*

39. Douglas Pimlott, Dougald Brown, and Kenneth Sam, *Oil under Ice* (Ottawa: CARC, 1976), especially chap. 8.

they cannot be handled effectively by states acting unilaterally within arbitrarily demarcated jurisdictional zones.[40] Arctic haze pervades the Arctic Basin; the waters of the Beaufort Sea wash back and forth across the Canadian-American boundary; and so forth.[41] Efforts to cope with these problems will require cooperative and coordinated actions on the part of all the Arctic states. Under the circumstances, the precedent that would be set by the resolution of issues relating to Arctic shipping through the demarcation of national jurisdictional zones would inevitably make it more difficult to solve many of the environmental concerns of the future in the Arctic.

Finally, the propensity to formulate issues in jurisdictional terms reinforces inappropriate forms of reasoning in the realm of Arctic shipping. Jurisdictional issues can only be resolved through a consistent application of the principles of either the public order of terrestrial spaces or the public order of nonterrestrial spaces. Faced with this choice, most analysts have endeavored to subsume Arctic issues under the principles of the evolving law of the sea.[42] But this practice regularly yields disturbing anomalies. The Arctic Basin is an ocean, but much of it has a permanent cover of pack or sea ice and the remainder is ice-covered for much of the year. The coastal littorals of the Arctic Basin, by contrast, are underlain with permafrost so that a great portion of this "land" actually consists of ice. As the Inuit residents of the Arctic have long known, these conditions make the ice of the Arctic Basin a suitable platform on which to carry out many activities that are ordinarily thought of as land-based activities, whereas the coastal littorals do not constitute a suitable platform for many traditional land-based activities.[43] It follows that efforts to force the Arctic into one or another of the major systems of public order must ultimately fail. They cannot resolve issues relating to Arctic shipping in an unambiguous fashion, and they will undoubtedly serve to hinder initiatives aimed at handling other Arctic issues in a constructive manner.

Even among those who do not question the proposition that Arctic issues should be approached in terms of the principles of the evolving law of the sea, the propensity to formulate these issues in jurisdictional terms leads to profound problems. Assume for a moment (though this is

40. For a more extensive analysis see Oran R. Young, *Resource Management at the International Level: The Case of the North Pacific* (London and New York: Pinter, 1977).

41. For more details consult Cynthia Lamson and David VanderZwaag, "Arctic Waters: Needs and Options for Canadian-American Cooperation," *Ocean Development and International Law* 18 (1987), 49–99.

42. Pharand, *Law of the Sea,* and Bloomfield, "Arctic."

43. Cooke and Van Alstine, *Sikumiut.*

certainly subject to dispute) that the 1982 Law of the Sea Convention is the appropriate place to look for authoritative formulations of the principles of the law of the sea.[44] With respect to the issues raised by Arctic shipping, the most striking feature of the convention is the loosely textured language of the relevant provisions.[45] How exactly should we evaluate claims regarding internal waters (Article 8)? Is Canada justified in drawing straight baselines around the entire Arctic Archipelago (Article 7)? Is the Northwest Passage an international strait or could it become one in the future (Article 34)? Would the regulatory authority accorded to coastal states in ice-covered areas (Article 234) supersede the transit rights of other states in international straits (Article 38)?[46] These and other similar questions cannot be resolved in an unambiguous fashion from a reading of the convention and associated documents. Plausible cases for any of a number of interpretations can be constructed in these terms.

Such ambiguities also occur regularly in municipal systems, but they are handled through a series of authoritative rulings handed down by appropriate courts. In the case of the outstanding issues relating to Arctic shipping, this approach to clarification is not likely to work. At the time of the enactment of the Arctic Waters Pollution Prevention Act of 1970, Canada announced that it would not accept the jurisdiction of the International Court of Justice over matters pertaining to the waters of the Northwest Passage and the Arctic Archipelago.[47] Nor is there any reason to expect the United States to be enthusiastic about this approach to resolving jurisdictional issues in the Arctic. What this means is that formulating issues relating to Arctic shipping in jurisdictional terms is apt to lead to a dead end legally. It can only complicate distributive bargaining between Canada and the United States with inconclusive legal rhetoric in contrast to facilitating the adoption of a problem-solving approach to common concerns.

What makes the present argument concerning the problems of juris-

44. For a discussion that raises doubts about the authoritativeness of the 1982 Law of the Sea Convention see James L. Malone, "Who Needs the Sea Treaty?" *Foreign Policy* 54 (1984), 44–63.

45. For the full text see United Nations, "United Nations Convention on the Law of the Sea," A/Conf.62/122 (1982).

46. For a provocative Canadian discussion of this issue see D. M. McRae and D. J. Goundry, "Environmental Jurisdiction in Arctic Waters: The Extent of Article 234," *University of British Columbia Law Review* 16 (1982), 197–228.

47. Dosman, *Arctic in Question,* and Richard B. Bilder, "The Canadian Arctic Waters Pollution Prevention Act: New Stresses on the Law of the Sea," *Michigan Law Review* 69 (1970), 1–54. More recently, the Canadian government has expressed a willingness to consider resorting to the International Court of Justice to resolve outstanding issues relating to Arctic waters.

dictional politics particularly apposite is the fact that there is an attractive method of setting aside jurisdictional issues relating to Arctic shipping without prejudice to the claims of either Canada or the United States. The simplest way to achieve this result would be to adapt the formula articulated in Article IV of the Antarctic Treaty of 1959 to apply to all areas that may be affected by Arctic shipping. This formula contains two distinct elements, both of which are important in the case of Arctic shipping. Article IV (1) states that agreement to set aside jurisdictional claims does not constitute a renunciation or diminution of any previously asserted rights or claims on the part of any of the contracting parties. Article IV (2) then enunciates a complementary commitment to the effect that no activities taking place under the Antarctic Treaty regime can form a basis for asserting, supporting, or denying the jurisdictional claims of the contracting parties. The effect of this formula is to leave all preexisting claims intact while allowing cooperative activities to take place in such a way that they will not have any impact on the future status of jurisdictional claims.[48] This formula says nothing about the framework for organizing cooperative activities in Antarctica. But it does serve to clear away an otherwise intractable complex of conflicting jurisdictional claims so that opportunities for cooperative ventures can be addressed in a spirit of integrative bargaining and joint problem solving.

AN ARCTIC SHIPPING REGIME

Should Canada and the United States agree to set aside jurisdictional claims relating to Arctic shipping, what alternative approach could they adopt? The answer proposed in this chapter is that Arctic shipping is an appropriate focus for regime formation at the present time.

Efforts to form mutually beneficial regimes can provide an effective method of resolving otherwise intractable disputes. Disputes relating to regional seas (a category to which the Arctic Ocean belongs) appear particularly susceptible to this approach to resolution.[49] The Convention on the Protection of the Marine Environment of the Baltic Sea, signed at Helsinki in 1974, promotes cooperative activities among a group of central and eastern European states that includes both East

48. F. M. Auburn, *Antarctic Law and Politics* (Bloomington: Indiana University Press, 1982).
49. See also Christopher B. Joyner, "Oceanic Pollution and the Southern Ocean: Rethinking the International Legal Implication for Antarctica," *Natural Resources Journal* 24 (1984), 1–40.

and West Germany as well as the Soviet Union. The 1976 Convention for the Protection of the Mediterranean Sea against Pollution (the Barcelona Convention) calls for coordinated action to control pollution among eighteen states that are divided by the East-West conflict, the Arab-Israeli conflict, and the Greek-Turkish conflict. Comparable efforts are now underway to enhance cooperation among the states bordering on the Caribbean Sea. The United Nations Environment Programme has backstopped several of these developments in a capable fashion through the articulation of its regional-seas program. Though no two cases are identical or even strictly comparable, it seems reasonable to infer from these experiences that the idea of forming an Arctic shipping regime is by no means farfetched.

Approaching issues relating to Arctic shipping as a problem of regime formation would yield substantial advantages over formulating them in jurisdictional terms. Working on the provisions of a coherent regime fosters a search for new options likely to yield mutual benefits rather than a preoccupation with dividing a predetermined payoff space or, in other words, a fixed pie.[50] This leads to the ascendance of integrative bargaining, in which the parties work together to enlarge the payoff-possibility set, over distributive bargaining. Likewise, the search for new options promotes attitudes of problem solving in contrast to an emphasis on devising winning tactics.[51] It becomes more important to think about the relative merits of various joint pollution-control programs, for example, than to back the other side into a tight spot in order to wring concessions on the demarcation of jurisdictional boundaries. Equally important, the integrative process involved in hammering out a mutually beneficial regime will ordinarily leave the parties with a sense of ownership of the product. Parties are much more likely to become committed to an arrangement that they have worked out together in a spirit of cooperation than to some division of a fixed pie that they feel pressured into accepting in nominal terms. This sense of ownership is likely to play a crucial role in facilitating implementation of resulting agreements and in promoting compliance among those subject to the rights and rules of regimes.

What can we say about the substantive content of a regime for Arctic shipping? It is pointless to attempt in this chapter to lay out a detailed blueprint for such a regime. There is a large family of possible institu-

50. For some imaginative suggestions regarding new types of international regimes see Finn Sollie et al., *The Challenge of New Territories* (Oslo: Universitetsforlaget, 1974).
51. Anatol Rapoport, "Strategic and Non-Strategic Approaches to Problems of Security and Peace," in Kathleen Archibald, ed., *Strategic Interaction and Conflict* (Berkeley: Institute of International Relations, 1966).

tional arrangements for Arctic shipping. The process of negotiating the terms of such a regime should lead to the identification of new options that none of us can foresee at this time. Also, it is important for the negotiating teams to formulate many of the provisions of an Arctic shipping regime on their own so that they will develop strong feelings of ownership regarding the outcome. Nonetheless, it is possible to identify what the principal components of an Arctic shipping regime should be and to spell out some of the basic options with regard to each of these components.[52]

General Character

Should an Arctic shipping regime be laid out in a formal treaty or would it suffice to let this regime take the form of a series of informal understandings among the participants? Those who adopt the perspectives of rational planning and institutional design will no doubt prefer to see the terms of such a regime spelled out formally. But it is worth noting that many social institutions operate effectively in the absence of formalization. Similarly, should an Arctic shipping regime involve some centralized organizational arrangements or would it be sufficient to settle for an agreement on the part of the members to coordinate their national policies and regulations on a decentralized basis? Again, centralized administrative arrangements appear to offer a more rational approach to management in an issue area like Arctic shipping. Given the character of international society, however, it may well be that decentralized coordination will yield more effective results than some form of centralized organization.

Coverage

What should be the coverage of an Arctic shipping regime in terms of geographical domain, functional scope, and membership?[53] There are compelling political as well as managerial reasons to extend the geographical domain of such a regime beyond the waters of the Canadian portion of the Northwest Passage and the adjacent Arctic Archipelago. Politically, it is desirable to devise a regime whose rights and rules would apply uniformly to all the North American Arctic states. This would foster a sense of equality and equity among the participants. In man-

52. For a parallel discussion of the international seabed regime see chapter 5 above.
53. See also Oran R. Young, *Resource Regimes: Natural Resources and Social Institutions* (Berkeley: University of California Press, 1982), especially chap. 4.

agerial terms, only a relatively broad domain would be sufficient to handle the problems posed by Arctic shipping. At a minimum, such a regime should cover the Beaufort Sea, Baffin Bay, and the Davis Strait in addition to the Canadian portion of the Northwest Passage; it should probably encompass segments of the Arctic Basin proper as well.

There is an equally strong case for endowing an Arctic shipping regime with a relatively broad functional scope. Arctic shipping clearly raises fundamental issues relating to pollution control. Such shipping also poses unavoidable questions regarding the maintenance of Arctic ecosystems and the viability of the socioeconomic and cultural systems of the human communities of the high Arctic. Efforts to address all these questions will require substantially expanded programs of research. Therefore, while there may well be political pressures to restrict the functional scope of an Arctic shipping regime (for example, confining it to relatively technical considerations pertaining to shipping per se), it would be unfortunate if these pressures were allowed simply to dictate the terms of a final agreement.

The issue of membership in an Arctic shipping regime is apt to be a controversial one. Should membership be limited to the littoral states (that is, Canada, Denmark/Greenland, and the United States)? Should other states wishing to engage in Arctic shipping (for example, Japan) be included and accorded equal status? What about various groups with legitimate interests in Arctic shipping (for example, the Inuit residents of the high Arctic, shipping companies, consumer groups, or environmental groups)? Is there a danger that the regime will look like an exclusive club to the remainder of the international community? These are all delicate questions that do not have any clearly correct answers.

Basic Principles

Underlying every regime or institutional arrangement are some basic principles that set the tone of the resultant practice. A few examples pertinent to the case of Arctic shipping will serve to clarify this proposition. The principle of setting aside jurisdictional claims without prejudice is obviously central to the case for an Arctic shipping regime as set forth in this chapter. The essential idea motivating this plan is to make use of the enterprise of institution building to transcend the politics of jurisdiction. But there is a strong case for the inclusion of several other principles as well. All legitimate interests (including those of the Inuit residents) should be fairly represented in the decision-making processes of the regime. Not only is this principle important from the point of view of justice, it is also central to the achievement of compliance with the provisions of the resultant practice. Participants should accord equal status to the experiential knowledge of the long-term residents of the high Arctic and to the scientific knowledge of Western researchers. It is essential to forge a strong partnership between ethnoscience and Western science in managing Arctic ecosystems.[54] Those involved in the regime should make every effort to arrive at major decisions through some consensual process rather than through exercises in coalition building or coercive diplomacy. This is just as important to reconciling the divergent interests of indigenous peoples and shipping companies as it is to satisfying the sensitivities of sovereign states such as Canada and the United States.

Substantive Provisions

The core of any coherent regime must be a system of rights and rules governing the specific actions of the participants.[55] In the case of Arctic shipping, these substantive provisions would involve such matters as construction standards for tankers, rules pertaining to safe operation under Arctic conditions, traffic control, aids to navigation (for example, search and rescue, ice-breaking assistance, ice-forecasting services), user fees, environmental protection, the socioeconomic integrity of nearby communities, liability rules for oil spills or other damages, and cleanup procedures. Though some of these matters are relatively technical, they form the substantive core of any management system for Arctic shipping and they deserve to be examined with great care. What is more, many of these matters offer considerable scope for the development of cooperative arrangements and the deployment of problem-solving techniques. It is worth noting the remarkable level of cooperation that already exists between Canada and the United States in several of these areas, despite the antagonism arising from jurisdictional differences.[56] This suggests that efforts to hammer out the substantive provisions of an Arctic shipping regime could become an important focus for integrative bargaining and the cultivation of problem-solving attitudes among those likely to become members of the regime.

Organizations

The distinction between institutional arrangements and explicit organizations is worth emphasizing in connection with Arctic shipping. The

54. See also William B. Kemp and Lorraine Brooks, "A New Approach to Northern Science," *The Northern Raven* 2, no. 1 (1983).
55. For a substantive case study involving the Arctic see Willy Ostreng, *Politics in High Latitudes: The Svalbard Archipelago* (London: C. Hurst, 1977).
56. For details see the essays in Dosman, *Arctic in Question.*

key issue here concerns the establishment of suitable organizations, if any, to administer an Arctic shipping regime. To lend substance to this concern, consider the following relevant models: the simple consultative arrangement of the Antarctic Treaty System, the commission established under the provisions of the international regime for whaling, and the more elaborate International Seabed Authority envisioned in connection with the seabed regime.[57] While it would be inappropriate to prescribe organizational mechanisms for an Arctic shipping regime at this stage, it seems likely that such a regime would require something more than the consultative arrangement of the Antarctic regime but considerably less than the proposed International Seabed Authority. Some sort of Arctic shipping commission including a staff of specialists possessing expertise in such matters as traffic control, ice forecasting, Arctic ecosystems, and socioeconomic impacts would probably suffice.

CONCLUSION

The thesis of this chapter is that the Canadian-American conflict over Arctic shipping is more apparent than real. It is, in essence, an artifact of the way in which issues relating to Arctic shipping are formulated. So long as these issues are posed in jurisdictional terms, the two countries are unlikely to be able to arrive at a comprehensive and mutually satisfactory agreement regarding Arctic shipping. If the jurisdictional problems are set aside without prejudice and issues relating to Arctic shipping are approached as problems of regime formation, on the other hand, numerous opportunities to pursue mutually beneficial processes of integrative bargaining will arise. In this regard, the Arctic shipping case exemplifies clearly the significance of alternative formulations of the basic issues in connection with social conflict more generally.

Despite the existence of opportunities to resolve issues relating to Arctic shipping through processes of integrative bargaining, the intermittent abrasiveness of Canadian-American relations in recent years may constitute a barrier to seizing these opportunities. Both sides have contributed to this state of affairs. Canada's economic nationalism, expansive attitudes toward marine jurisdiction, and declining interest in defense have undoubtedly irritated American policymakers. By the same token, the unwillingness of the United States to pay serious attention to Canada's concerns about acid precipitation, fisheries management, and economic autonomy has provoked negative feelings toward the United States among Canadian policymakers. Approached from a

57. See also chapter 5 above.

different angle, however, a constructive resolution of Canadian-American differences over Arctic shipping could become a catalyst for an era of more harmonious relations between the two countries. The new governments expected to take office in both the United States and Canada in 1989 might therefore do well to seize on Arctic shipping as an issue area in which constructive steps of considerable symbolic significance could be taken with relative ease.

PART THREE

REGIME ANALYSIS: PRESENT STATUS AND FUTURE PROSPECTS

Prologue

Transboundary environmental issues, ranging from the localized concerns of two countries sharing a river basin to the global concerns associated with ozone depletion or the buildup of carbon dioxide, have arrived on the international agenda to stay. If anything, these issues will move closer to center stage in international society during the years to come. But what can we conclude about the value of regime analysis in enhancing our understanding of these issues? Is the research program embedded in this line of thinking likely to become a continuing feature of the strategy we employ in our studies of international relations? Chapter 8 argues, in essence, that the jury is still out on these questions and that it will remain out until we reach a clearer understanding of the extent to which institutional arrangements operate as determinants of collective outcomes at the international level. The pursuit of this goal, in turn, will require a reexamination and refinement of the theoretical models we employ in thinking about the behavior of the individual actors in international society. Accordingly, it may well be some time before it is possible to judge accurately whether regime analysis will fade away in the manner of other fads in the field of international relations or flourish as an important part of the analytic apparatus of this field of study.

All students of international affairs share an interest in deepening and broadening our comprehension of recurrent international phenomena. Understandably, however, some analysts along with most policymakers want to go a step further to explore the implications of any new approach to understanding, such as regime analysis, in applied or policy-relevant terms. This concern comes down to the question of whether regime analysis has much to offer those engaged in institutional design in international society. Can students of international regimes bring their knowledge of institutional arrangements to bear in such a way as to help those struggling to devise workable regimes to govern human activities affecting various marine mammals, deep-seabed minerals, the electromagnetic spectrum, or the stratospheric ozone layer? Chapter 9 turns to a discussion of this question. The chapter develops the argument that the relationship between regime analysis and institutional design differs from the more well-known relationship between theory and application in fields such as medicine and engineering. Efforts to design institutional arrangements at the international level always involve complex interactions among a number of self-interested actors endeavoring to promote their own ends. It follows that the essential problem of institutional design in international society is to guide such interactions toward the acceptance of preferred arrangements rather than simply to prescribe appropriate institutional arrangements on the assumption that some rational "lawgiver" will naturally follow these prescriptions. As every well-trained student of international relations knows, there is nothing automatic about the adoption of socially desirable or optimal regimes in the interactions of those involved in regime formation at the international level. At the same time, we must guard against becoming overly pessimistic about our ability to guide these interactions toward constructive outcomes. Following an examination of the constraints on institutional design in international society, therefore, chapter 9 outlines a strategy for those wishing to maximize the probability of arriving at agreements, whether explicit or implicit, on appropriate regimes for natural resources and the environment as well as other pressing international concerns.

CHAPTER EIGHT

Analysis: Toward a New Theory
of International Institutions

Current events play a determinative role in setting research agendas in the field of international relations. During the height of the Cold War in the 1950s and 1960s, scholars rushed to improve our understanding of strategic interaction and conflict. Following some years of detente, the souring of Soviet-American relations during the late 1970s and early 1980s provided a new stimulus for studies of international conflict. A wave of crises after World War II, especially the more anxiety-producing confrontations such as the Berlin crisis of 1961 and the Cuban missile crisis of 1962, triggered a burst of academic interest in the generic phenomenon of crisis. The emergence of the European Economic Community during the 1960s sparked a growth industry among those seeking to analyze functional integration and to construct neo-functional theories of economic and political integration. The increasing prominence of multinational corporations during the 1970s spurred students of international relations to break away from state-centered perspectives and to investigate the nature of transnational relations. Similarly, the recent explosion of evidence regarding such environmental problems as the greenhouse effect, ozone depletion, and the long-range transport of acid deposition has produced a sharp rise of interest in questions relating to the management of the global commons. There is nothing abnormal about such a relationship between current events and research programs; it occurs in many fields to one degree or another. In the field of international relations, however, the result has been a succession of transient fads. As Strange puts it, the outcome all too often is "one of those shifts of fashion not too difficult to explain as a temporary reaction to events in the real world but in itself making little in the way of a long-term contribution to knowledge."[1]

Without doubt, the current burst of work on regimes, or, more broadly, international institutions, reflects an emerging sense, especially among Americans, that the international order engineered by the United States and its allies in the aftermath of World War II is eroding rapidly and may even be approaching the verge of collapse. The international-trade regime seems incapable of stemming a mounting tide of protectionist pressures from many corners of the world, including the United States itself. The monetary regime is staggering under the impact of a series of debt crises and may go under as these crises become increasingly severe over the next few years. Accordingly, the surge of interest in international institutions, a subject decidedly out of fashion during the 1970s, is easy enough to explain as a response to current events. But is the resultant flow of scholarly work on international regimes any more likely to yield lasting contributions to knowledge than other recent fads or fashions in the field of international relations?

As I suggested in the prologue to Part 3, the jury is still out on this issue. Even so, the literature on regimes has introduced a conception of international institutions that differs markedly from the conception embedded in the orthodox literature on international relations in general and international organizations in particular. In my judgment, this conception has already helped to identify constructive opportunities for reintegrating the subfields of international politics, economics, law, and organization.[2]

WHAT ARE INTERNATIONAL REGIMES?

A number of those writing about institutional arrangements in international society have come to rely on a common definition of the concept of regimes. In the well-known formulation of Krasner, regimes are "sets of implicit or explicit principles, norms, rules, and decision-making procedures around which actors' expectations converge in a given area of international relations."[3] This apparent definitional consensus is a

1. Susan Strange, "*Cave! hic dragones:* A Critique of Regime Analysis," in Stephen D. Krasner, ed., *International Regimes* (Ithaca: Cornell University Press, 1983), 337.

2. See also the helpful commentary on the regimes literature in Stephan Haggard and Beth A. Simmons, "Theories of International Regimes," *International Organization* 41 (1987), 491–517.

3. Stephen D. Krasner, "Structural Causes and Regime Consequences: Regimes as Intervening Variables," in Krasner, *International Regimes*, 2.

remarkable achievement in a field of study that is as intellectually anar-chical as international relations has been over the last generation. But it is not sufficient to suppress insistent criticisms regarding the clarity and, therefore, the utility of the concept of international regimes. Strange, for example, states flatly that the notion of regimes "is yet one more woolly concept that is a fertile source of discussion simply because people mean different things when they use it."[4] Another prominent commentator, speaking for many readers of the regimes literature, has recently characterized the concept of a regime as "vague" and argued that it "cries out for conceptual development."[5] Accordingly, it seems pointless to proceed to an assessment of substantive issues raised by the regimes literature without some consideration of the conceptual prob-lems plaguing the notion of regimes.

Part of the problem with the definition that Krasner sets forth is that it does not allow us to identify regimes with precision and to separate regimes easily from the rest of international relations. The common definition is really only a list of elements that are hard to differentiate conceptually and that often overlap in real-world situations. In an ef-fort to address this problem, Krasner offers the following elaboration: "principles are beliefs of fact, causation, and rectitude. Norms are stan-dards of behavior defined in terms of rights and obligations. Rules are specific prescriptions or proscriptions for action. Decision-making pro-cedures are prevailing practices for making and implementing collec-tive choice."[6] But now we must cope with another set of ambiguous terms in the form of beliefs, standards, prescriptions, and practices in addition to the original set consisting of principles, norms, rules, and procedures. As Kratochwil and others have demonstrated, there are various ways to explicate concepts such as rules, prescriptions, norms, and practices, and choices made at this level exert a profound impact on subsequent analyses.[7]

The common definition also exhibits a disconcerting elasticity when applied to the real world of international relations. As Strange observes, some writers adopt a narrow view, using the concept of a regime to apply to "internationally agreed arrangements, usually executed with the help

4. Strange, *"Cave! hic dragones,"* 342–343.

5. Friedrich Kratochwil, "The Force of Prescriptions," *International Organization* 38 (1984), 685.

6. Krasner, "Structural Causes and Regime Consequences," 2.

7. Kratochwil, "Force of Prescriptions," 685–703. For a suggestive account that em-phasizes the idea of operationalizing norms in behavioral terms, however, see Robert Axelrod, "An Evolutionary Approach to Norms," *American Political Science Review* 80 (1986), 1095–1111.

of an international organization."[8] Others, she points out, broaden the concept to encompass "almost any fairly stable distribution of power to influence outcomes," thereby blurring the distinction between international regimes and the entire realm of international relations.[9] Under the circumstances, Strange is surely justified in questioning whether there is going to be "much useful or substantial convergence of conclusions about the answers to . . . questions concerning [the] making and unmaking" of international regimes.[10]

More generally, the notion of regimes employed in much of the recent literature is conceptually thin. It provides a checklist of elements whose occurrence in the same time and space is sufficient to place a phenomenon in the category labeled "regimes." But it does not tie the concept into any larger system of ideas that would help to solve the definitional ambiguities discussed in the preceding paragraphs and that would offer guidance in formulating key questions and hypotheses regarding international regimes. Certainly, it is wrong to expect too much of definitions. Especially in efforts to come to grips with intangible phenomena, however, definitions that are not clearly integrated into some larger conceptual system inevitably lead to trouble. Thus, Olson's concept of collective goods, Rawls's concept of justice as fairness, or Raiffa's concept of a zone of agreement would all be virtually useless if they were not embedded in larger systems of ideas serving to flesh them out and to guide thinking about their use in understanding the world of observable events.[11] More than anything else, it is this problem of conceptual thinness that could relegate the study of international regimes to the status of a passing fashion in contrast to a long-term contribution to knowledge.

To ensure that it does yield a long-term contribution to knowledge, I believe, such work should start with the proposition that international regimes are social institutions and tie the analysis of regimes directly to the study of institutions more generally. Social institutions are behaviorally recognizable practices consisting of roles linked together by clusters of rules or conventions governing relations among the occupants of these roles. It is not difficult to apply this conception of institutions to the particular case of international regimes. As the preceding chapters

8. Strange, *"Cave! hic dragones,"* 343.
9. Ibid.
10. Ibid.
11. Mancur Olson, Jr., *The Logic of Collective Action* (Cambridge: Harvard University Press, 1965); John Rawls, *A Theory of Justice* (Cambridge: Harvard University Press, 1971); Howard Raiffa, *The Art and Science of Negotiation* (Cambridge: Harvard University Press, 1982).

have demonstrated, international society is well-stocked with regimes dealing not only with trade and monetary relations but also with many other activities such as high-seas fishing, the use of the electromagnetic spectrum, deep-seabed mining, and the problem of nuclear accidents.

It is tempting to join writers like Rawls in approaching social institutions as practices that participants select in a conscious and explicit fashion.[12] Some of those who have contributed to the literature on international regimes often seem to proceed in this way.[13] Though this approach has the attraction of making it seemingly easy to identify regimes empirically, it is ultimately misleading. As many observers of social institutions have noticed, such arrangements almost always encompass informal elements and they sometimes arise through self-generating processes in which the participants are not at all conscious that they are engaged in the formation of social institutions.[14] What is more, social institutions commonly acquire a remarkably coercive quality. Actors are regularly socialized into accepting and performing the roles associated with institutional arrangements, and the costs to individual actors of opting out of participation in prevailing institutions are often prohibitive. This is undoubtedly a major factor accounting for the longevity of many social institutions, even those that become severe threats to social welfare, such as warfare in the nuclear age.

Yet it would be wrong to suppose that social institutions are static or unchanging. The political, economic, technological, and moral conditions underpinning institutions sometimes shift in such a way that the institutions themselves atrophy (for example, colonialism in international society). In other cases, circumstances converge to produce an abrupt upheaval in which many prevailing social institutions are swept aside (for example, the ancien régime in the French Revolution). Note, however, that changes of this sort do not provide any basis for optimism for those whose interest in regimes arises fundamentally from a desire to engage in institutional design or social engineering. The fact that the trade and monetary regimes established in the aftermath of World War II may now be on the verge of collapse offers no assurance that those

12. See Rawls, *A Theory of Justice*, for an account in which practices are analyzed in contractarian terms.

13. See, for example, Robert O. Keohane, "The Demand for International Regimes," 141–171 in Krasner, *International Regimes*, and Arthur A. Stein, "Coordination and Collaboration: Regimes in an Anarchic World," 115–140 in Krasner, *International Regimes*.

14. So, for example, confining social institutions to practices that participants select in a conscious fashion would exclude an array of markets or exchange systems that many observers regard as paradigmatic examples of institutional arrangements. For a more extended account of the emergence of practices through self-generating processes see Robert Axelrod, *The Evolution of Cooperation* (New York: Basic Books, 1984).

likely to be affected by such a collapse will be able or even willing to restructure these regimes on a planned basis.

Of equal importance for this examination of the burgeoning literature on international regimes is the distinction between social institutions and organizations. Institutions are behaviorally recognizable practices made up of sets of roles coupled with clusters of rules or conventions governing relations among the occupants of these roles. Organizations are physical entities possessing offices, personnel, equipment, budgets, and so forth. Both the International Monetary Fund and the World Bank are organizations. The Bretton Woods system, by contrast, refers to an international institution or regime.[15] A striking feature of the postwar international-trade regime is that the GATT establishes an institutional arrangement requiring little in the way of organizations, whereas the rejected International Trade Organization would have coupled the trade regime with an explicit organization.[16] A common mistake among commentators on international relations is to say that international society features few effective institutions when, in fact, they mean to assert that it has few effective international organizations. There is no simple relationship between institutions or regimes and organizations. Some social institutions (for example, open-to-entry common-property regimes during periods of light usage) operate effectively in the absence of any elaborate organizations.[17] Organizations sometimes come into existence in the absence of any well-established counterpart institutions (for example, the United Nations Organization after World War II). The relationships that can and do arise between regimes and organizations should therefore be a subject of sustained interest to students of international relations.

The Origins of Regimes

To enquire into the origins of regimes is to ask both why social institutions arise in international society and how such institutions actually emerge in a social setting that is generally regarded as anarchical.[18]

---

15. Benjamin J. Cohen, "Balance-of-Payments Financing: Evolution of a Regime," 315–336 in Krasner, *International Regimes*.

16. On the GATT system, see Jock A. Finlayson and Mark W. Zacher, "The GATT and the Regulation of Trade Barriers: Regime Dynamics and Functions," 273–314 in Krasner, *International Regimes*.

17. A case in point is the regime for high-seas fishing prior to the dramatic advances of the postwar era in the technology of fishing. See Francis T. Christy and Anthony Scott, *The Common Wealth of Ocean Fishing* (Baltimore: Johns Hopkins University Press, 1965).

18. For a well-known description of international society in these terms see Hedley Bull,

Two features of the emerging literature on international regimes stand out in this context. There is broad agreement concerning the reasons why the members of the international community allow themselves to be drawn into regimes or social institutions. Yet there is an equal amount of confusion and disagreement about the actual processes through which regimes come into existence.

By now, everyone is aware that rational egoists operating in the absence of effective rules or social conventions often fail to realize feasible joint gains and end up with outcomes that are suboptimal (sometimes dramatically suboptimal) for all concerned.[19] Perhaps the classic exemplar for these problems of collective action is the prisoner's dilemma, a paradigm that many observers have scrutinized closely and applied repeatedly to the world of international relations. But the same basic conclusion emerges from theoretical studies of the provision of collective goods, common-property difficulties, social traps, and security dilemmas as well as from empirical studies of protectionist measures, trade wars, arms races, and so forth. It follows that individual actors frequently experience powerful incentives to accept behavioral constraints of the sort associated with institutional arrangements in order to maximize their own long-term gains, regardless of their attitudes toward the common good. Whether individual actors approach such issues in terms of rule utilitarianism, the tenets of an ethical system, or some sort of nonutilitarian contractarianism is not critical at this juncture. The point is that it is easy to comprehend why actors would willingly abandon a truly anarchical social environment for a world featuring recognizable social institutions, though this need not imply a willingness to accept any elaborate organizations (for example, a world government).

Numerous writers have developed this story in a clear and convincing manner. Among those currently interested in international regimes, however, Keohane has provided the most extensive and, in some respects, the most peculiar treatment of this subject.[20] He analyzes regime formation at length in terms of a vocabulary adapted from microeconomics. Thus, the development of regimes constitutes a response to "political market failure" on the part of actors who assess their options in terms of a kind of rule utilitarianism and who employ reasonably low discount rates in computing the present value of future benefits. It is an

The Anarchical Society: A Study of Order in World Politics (New York: Columbia University Press, 1977).

19. For a sophisticated review of these problems of "collective action" see Russell Hardin, *Collective Action* (Baltimore: Johns Hopkins University Press, 1982).

20. Robert O. Keohane, *After Hegemony: Cooperation and Discord in the World Political Economy* (Princeton: Princeton University Press, 1984), 49–132.

easy step from this starting point to a consideration of the "demand" for international regimes as well as to an analysis in which actors deliberately enter into institutional arrangements through some sort of bargaining process so long as the marginal benefits from doing so outweigh the marginal costs.[21] I have no serious quarrel with Keohane's choice of this vocabulary in his effort to explain why individual actors experience powerful incentives to enter into social institutions that subsequently become significant constraints on their freedom of action. As I shall demonstrate in a subsequent section of this chapter, however, the use of this microeconomic vocabulary does introduce profound biases into any consideration of the significance of international regimes.

When we turn to the actual processes through which institutions emerge, by contrast, confusion abounds. Many writers (for example, Keohane and Stein) seem to suggest that regime formation is a process of explicit bargaining in which two or more participants, perceiving a zone of agreement or contract zone, negotiate openly until they reach agreement on the provisions of mutually satisfactory institutional arrangements, which can be cast in the form of a treaty or convention. In fact, these writers are prone to assert that arrangements emerging in any other way do not really qualify as regimes.[22] This has the effect of directing the attention of those interested in international regimes almost exclusively toward institutional arrangements that are made explicit or formalized in contracts or treaties (for example, the postwar international-trade and monetary regimes). And it implies that the study of regime formation is merely a special case of the analysis of explicit bargaining.

At the same time, however, a number of students of international regimes (including Keohane himself) have devoted much time and energy to an examination of the role of hegemons or dominant actors in the formation of specific regimes.[23] Some have even suggested that the presence of a hegemon constitutes a necessary (though not sufficient) condition for success in regime formation. This inevitably leads to confusion regarding the dynamics of regime formation. There is nothing in theories of bargaining or negotiation as such to justify the conclusion that a hegemon is needed to produce agreement, so long as a contract zone or a zone of agreement exists. On the contrary, the usual assumption embedded in such theories is that rational actors will find a way to realize feasible joint gains. The concept of a hegemon, on the other

21. Ibid., 76.
22. Ibid.
23. Ibid., 135–181.

hand, is closely associated with the idea of an individual member of a group possessing the effective capacity or power to impose institutional arrangements on the group regardless of the preferences of the other members. Under the circumstances, while situations featuring bargaining are apt to yield what I have called negotiated arrangements, the presence of a true hegemon will lead to the emergence of imposed arrangements.

No doubt, many interactions regarding social institutions involve complex mixtures of conventional bargaining and the sort of coercion associated with the activities of unusually powerful actors. Surely, there is nothing uncommon about situations in which the relevant parties differ substantially in their bargaining strength and, therefore, their ability to influence the content of the provisions incorporated into international regimes. Under the circumstances, it probably makes sense to join Kindleberger in employing the notion of leadership to characterize the role of those who are particularly influential in shaping the content of social institutions.[24] Leadership in connection with regime formation is entirely compatible with processes of integrative bargaining. In the typical case, in fact, the leader will rely more on entrepreneurial skills than on the power to impose arrangements on others against their will. Still, it is undeniable that many efforts to devise institutional arrangements in international society involve a significant element of coercion, even (or perhaps especially) when no single participant qualifies for the role of hegemon. This combination has given rise to the concept of coercive diplomacy, a notion that reflects the complex mixture of negotiation and coercion present in many interactions at the international level.[25] Even so, negotiation and coercion are distinct processes; it does not clarify matters to conflate them in thinking about the origins of international regimes.

Note also that some of the most provocative recent theoretical work on the emergence of rules or social conventions among rational egoists focuses on spontaneous or, to use Hayek's term, self-generating institutional arrangements that arise in the absence of either conventional bargaining or coercive pressure. This includes the work of Lewis and

24. Thus, Kindleberger has proposed that we "think of leadership or responsibility" rather than hegemonic power. See Charles P. Kindleberger, "International Public Goods without International Government," *American Economic Review* 76 (1986), 10, and "Dominance and Leadership in the International Economy," *International Studies Quarterly* 25 (1981), 242–254.

25. On coercive diplomacy, consult Thomas C. Schelling, *Arms and Influence* (New Haven: Yale University Press, 1966), and Oran R. Young, *The Politics of Force: Bargaining during International Crises* (Princeton: Princeton University Press, 1968).

Hardin on social conventions as well as the work of Schelling on tacit bargaining and the role of k groups in solving collective-action problems.[26] Even more suggestive is the recent work of Axelrod (who has a long-standing interest in international relations) on the "evolution of cooperation."[27] This work demonstrates that purely self-interested actors may develop effective rules or social conventions through an interactive learning process involving trial and error coupled with a kind of behavioral natural selection, so long as they expect to interact with each other repeatedly and employ relatively low discount rates in computing the present value of future benefits.[28]

The emphasis on negotiated institutions that pervades much of the literature on international regimes may be a reflection of the liberal biases of the authors. It is primarily in the context of such a contractarian perspective, after all, that it makes sense to think seriously about the prospects for institutional design regarding international regimes. Nonetheless, it is unfortunate that many of those who are interested in institutional arrangements in international society have chosen to ignore or minimize the significance of coercive pressures and, especially, spontaneous processes in thinking about the formation of regimes. In so doing, they inevitably produce one-dimensional accounts of the emergence of institutional arrangements in international society. To counter this bias, we should start our examination with clear analytic distinctions among negotiation, imposition, and spontaneous processes. We would then be in a position to produce a far richer account of regime formation, appraising systematically the conditions under which one or another of these processes is likely to predominate as well as the conditions under which two or more of them are likely to occur in some complex mixture.

STABILITY AND CHANGE

It is not an exaggeration to say that a search for determinants of stability and change in international institutions is the most striking feature of the recent regimes literature.[29] This is perhaps understand-

26. David K. Lewis, *Convention: A Philosophical Study* (Cambridge: Harvard University Press, 1969); Hardin, *Collective Action*; and Thomas C. Schelling, *Micromotives and Macrobehavior* (New York: Norton, 1978), chap. 7.

27. Axelrod, *Evolution of Cooperation*, and "An Evolutionary Approach to Norms."

28. For a thoughtful review that discusses some of the limits of Axelrod's analysis under real-world conditions, see Joanne Gowa, "Anarchy, Egoism, and Third Images: *The Evolution of Cooperation* and international relations," *International Organization* 40 (1986), 167–186.

29. Keohane, *After Hegemony*, 43.

able in the light of growing concerns about the future of some of the major postwar international institutions, such as the regimes for high-seas fishing and other global commons, not to mention the international trade and monetary regimes. Nor is there anything peculiar or inappropriate about a pattern in which the attention of scholars follows trends in real-world activities. Despite the efforts scholars have devoted to the analysis of stability and change in regimes, however, there is considerable confusion in the literature about this aspect of international regimes.

A debate focusing on what has come to be known as the hegemonic-stability hypothesis has dominated scholarship in this realm for a number of years. As Keohane (drawing on Kindleberger's earlier work on the Great Depression), characterizes it, this hypothesis suggests that the presence of a dominant party is critical (perhaps even necessary) not only to the initial formation of regimes but also to the maintenance of institutional arrangements over time.[30] It is easy enough to see how observers of the American role in the aftermath of World War II might have arrived at an hypothesis along these lines. Even so, the hypothesis does not withstand the test of analytic scrutiny. Its theoretical underpinnings are shaky at best. No doubt, the presence of a dominant player may give rise to an Olsonian privileged group.[31] Also, the exertion of coercive pressure on the part of a particularly powerful actor or an effective ruling coalition will be critical to any effort to form imposed regimes or institutional arrangements. But there is nothing in the theoretical literature on bargaining to suggest that social contracts will come unglued in the absence of a dominant partner. If anything, arguments emphasizing transaction costs and the problems of collective action suggest that parties will be hesitant to discard existing institutional arrangements, even when changing circumstances leave these arrangements poorly adapted to emerging social conditions. For their part, spontaneous institutional arrangements often exhibit great continuity, as well as great flexibility, even in large groups of "small" actors. And all social institutions tend to acquire considerable staying power of an inertial nature. The costs to individual actors of ignoring or breaking away from established institutions on a unilateral basis are generally high. Even leading members of groups experience great social pressure to conform, at least in a rough and ready way, to the rules or conventions of social institutions. What is more, the difficulties of putting together a

30. Ibid., 31. For a restatement of Kindleberger's views in which he emphasizes the idea of leadership and disassociates himself from the concept of hegemony see Kindleberger, "International Public Goods," 10.
31. Olson, *Logic of Collective Action*, chap. 1.

winning coalition, much less achieving general consensus, in support of specific alternatives to prevailing institutional arrangements are notorious.

Nor do empirical studies of regimes support the hegemonic-stability hypothesis. To be sure, sharp shifts in the distribution of power can and often do lead, over time, to adjustments in existing social institutions. But as Keohane argues in a recent statement on the subject, "Hegemony and international regimes can both contribute to cooperation."[32] That is, the emergence of effective institutions can become a source of order or stability quite apart from the actions of a dominant power or even in the absence of such a power. The upshot of all this is that the hegemonic-stability hypothesis is dead. Keohane himself puts it well in saying that "the dominance of a single great power can contribute to order in world politics, in particular circumstances, but it is not a sufficient condition and there is little reason to believe that it is necessary."[33] Perhaps those earlier students of international politics who placed great emphasis on the existence of a balance of power in maintaining institutional arrangements in international society can be forgiven a sense of deja vu in reading a passage like this.

What are the implications of this discussion for the study of international regimes? It reveals, to begin with, the extent to which the recent literature on regimes is associated with the American study of international relations. The erosion of American dominance has progressed far enough to call into question some of the familiar assumptions underlying American foreign policy. Keohane is quite explicit about this in asking whether and how cooperation among the advanced capitalist countries can persist without the dominance of the United States.[34] The same underlying concern surely motivates many of the other contributors to the regimes literature. Yet this concern will undoubtedly seem both self-serving and ironic to many non-American students of international relations. As Strange (among others) argues, the United States continues to be an enormously powerful actor in the international system.[35] Further, many non-Americans (and even enlightened Americans) clearly regard the restoration of Europe and Japan to the status of major powers in the international community as a healthy development, rather than as a cause for concern. In the classic formulations (from Hobbes to Rawls), after all, the central puzzle associated with efforts to

32. Keohane, *After Hegemony*, 240.
33. Ibid., 46.
34. Ibid., 41–43.
35. Strange, *"Cave! hic dragones,"* 340–341, and "The Persistent Myth of Lost Hegemony," *International Organization* 41 (1987), 551–574.

devise workable social contracts or agreements regarding social institutions (though not necessarily organizations) is the search for mutually acceptable constraints on the actions of those who are, at least in some rough sense, equals.[36] It is not the perpetuation or the legitimation of dominance on the part of a hegemonic power.

The concern over the possible threat to international cooperation posed by the erosion of American hegemony is also ironic in light of the fact that the ideology of capitalism and free enterprise, at least in its American variant, touts the virtues of a world of numerous equals (that is, price takers in a system of exchange relations) over a world featuring a dominant power (that is, a monopolist or a monopsonist in market terms). Why should this supposition, so widely espoused by American observers at the domestic level, be inapplicable at the level of international society? It follows that commentators in other countries may not only regard the concern over regime stability in the absence of American hegemony as a peculiarly American problem, they may also find it hard to sympathize with the true motivation underlying this concern. To quote Strange again, "To non-American eyes . . . , there is something quite exaggerated in the weeping and wailing and wringing of American hands over the fall of the imperial republic."[37]

These comments aside, however, it is certainly worthwhile to conduct a more general enquiry into the conditions governing stability and change in international institutions. As a first step in such an enquiry, in my judgment, we need to replace Keohane's emphasis on "after hegemony," which directs attention to the problem of maintaining international institutions following the erosion of American dominance, with an emphasis on "beyond hegemony," which would encourage a more wide-ranging examination of the full set of conditions governing the rise and fall of institutional arrangements at the international level. Institutions change in response to an array of political, economic, technological, sociocultural, and even moral developments. Undoubtedly, institutional arrangements in all social systems do adjust to sharp shifts in the distribution of power. But consider also the role of changes in prevailing structures of property rights in giving rise to the European-dominated institutions of the international states system during the sixteenth and seventeenth centuries.[38] Think about the consequences of technological

36. Hobbes attaches particular importance to the rough equality of individuals in the state of nature because he assumes that virtually every individual in such a state will possess the physical ability to kill any other member of the group.

37. Strange, *"Cave! hic dragones,"* 340.

38. Douglass C. North and Robert Paul Thomas, *The Rise of the Western World* (Cambridge: Cambridge University Press, 1973).

developments, such as the impact of nuclear weapons on war as a social institution or the impact of high-endurance stern trawlers on open-to-entry common-property regimes for the marine fisheries.[39] And let us not forget the role of evolving ethical or moral sensibilities in delegitimizing the institution of colonialism during the twentieth century.[40]

I conclude from this that there are no necessary conditions for change in international regimes and that any of a variety of factors may be sufficient to precipitate major changes in prevailing social institutions in real-world situations. This may seem frustrating to those seeking to construct a parsimonious theory of stability and change in international regimes. For all its shortcomings, the hegemonic-stability hypothesis offered the attractions of a simple theory emphasizing a single, master variable. Yet the argument I have sketched here suggests that studies of stability and change can become a rich vein in the literature on international regimes. It also offers some guidelines for those wondering how to approach this subject following the demise of the hegemonic-stability hypothesis.

DO REGIMES MATTER?

One of the more surprising features of the emerging literature on regimes is the relative absence of sustained discussions of the significance of regimes, or, more broadly, social institutions, as determinants of collective outcomes at the international level. To be sure, Krasner presents a brief account of regimes as intervening variables.[41] Ruggie and his colleagues make some attempt to explore the place of international institutions in the "matrix of constraints and opportunities" that developing countries face in efforts to improve their lot.[42] Keohane obviously believes institutions can and do have an impact on state behavior.[43] But he devotes the bulk of his energy to a consideration of regime formation and change. The result is something of an analytic vacuum. The ultimate justification for devoting substantial time and energy to the study of regimes must be the proposition that we can account for a good deal of the variance in collective outcomes at the

39. William W. Warner, *Distant Water: The Fate of the North Atlantic Fishermen* (Boston: Little, Brown, 1983).

40. A. P. Thornton, *The Imperial Idea and Its Enemies* (New York: Macmillan, 1959).

41. Krasner, "Structural Causes and Regime Consequences," 5–10.

42. John Gerard Ruggie, ed., *The Antinomies of Interdependence* (New York: Columbia University Press, 1983), especially 462–465.

43. Keohane, *After Hegemony*, 26.

international level in terms of the impact of institutional arrangements. For the most part, however, this proposition is relegated to the realm of assumptions rather than brought to the forefront as a focus for analytical and empirical investigation.

To make matters worse, the assumptions that various groups of thinkers bring to bear on this subject differ sharply. Political scientists steeped in the power-oriented perspectives of realism or trained in the empirical methodologies of behavioralism tend to dismiss any emphasis on the role of institutions as a vestige of the discredited ideas of the formal, legal, institutional school of thought. Yet other students of politics as well as most lawyers (who typically make a living by devising, interpreting, and refining institutional arrangements) cannot imagine treating institutions as anything but central determinants of collective behavior. Mainstream economists, proud of the formalizations of neoclassical economics emerging in the wake of the dismissal of the older institutional economics of Veblen, Commons, and Mitchell, generally ignore institutional issues (via a liberal use of ceteris paribus assumptions) or regard them as hopelessly intractable. But the public-choice movement, as exemplified by the works of Arrow, Buchanan, Tullock, and Olson, places great weight on the character of institutional arrangements in explaining collective choices or macrobehavior.[44] For their part, many Marxists also stress the role of institutions (for example, the international division of labor) in analyzing issues like "imperialist" conflicts among capitalist states and the structural bases of dependency among the less developed countries.

Where does the truth lie regarding the significance of institutions as determinants of collective behavior? Almost everyone believes, in the final analysis, that social institutions are of great importance in domestic society. To think otherwise would make it necessary to grapple with profound questions about any number of propositions dealing with the relative merits of alternative electoral procedures, structures of property rights, economic systems, mechanisms for resolving disputes, and so forth. No doubt, it is tempting simply to transfer this conviction from the domestic realm to the international arena. But this will not do, since many influential writers assert that international society differs from domestic society precisely in ways that diminish the significance of institutions at the international level. They would agree with Strange that it is possible to understand who gets what in international society "by

44. Kenneth Arrow, *Social Choice and Individual Values*, 2d ed. (New York: Wiley, 1963); James Buchanan and Gordon Tullock, *The Calculus of Consent* (Ann Arbor: University of Michigan Press, 1962); and Olson, *Logic of Collective Action*.

looking not at the regime that emerges on the surface but underneath, at the bargains on which it is based."[45] In the final analysis, such analysts regard international regimes as epiphenomena whose dictates are apt to be ignored whenever actors find it inconvenient or costly to comply with them and whose substantive provisions are readily changeable whenever powerful members of the community find them cumbersome or otherwise outmoded.[46] What is more, they argue that it is perfectly possible to construct satisfactory explanations of collective behavior at the international level in terms of factors that have nothing to do with institutions (for example, prevailing configurations of power and interest or geopolitical considerations). From this perspective, the emphasis on institutions in the literature on international regimes not only encourages the articulation of unnecessarily cumbersome explanations it also diverts attention from the real determinants of collective behavior at the international level.

All this suggests the need for a more sustained effort to evaluate the significance of international regimes. In part, this effort must surely take the form of careful empirical work. Taking a cue from the public-choice movement, students of international regimes might well consider devising a series of controlled experiments to investigate the consequences of alternative institutional arrangements under conditions that simulate those prevailing in international society.[47] There is also room for natural experiments comparing and contrasting real-world regimes, both synchronically and diachronically, in an effort to isolate the role of institutional arrangements in shaping collective behavior. The trick here is to select cases, at least initially, in such a way as to hold as much as possible constant, other than the relevant institutional arrangements themselves. Such a procedure will minimize interpretive ambiguity and make it possible to attribute variance in collective outcomes to the impact of institutional arrangements with some degree of confidence. This observation suggests that the postwar trade and monetary regimes, which constitute the principal focus of the existing literature on regimes, are not good cases for a rigorous, empirical examination of the significance of international regimes. Because they arose in an environment in

45. Strange, "Cave! hic dragones," 354.
46. See also Strange's comment that "all those international arrangements dignified by the label regime are only too easily upset when either the balance of bargaining power or the perception of national interest (or both together) change among those states who negotiate them" (ibid., 345).
47. While Axelrod's recent work emphasizes behavioral parameters and the strategies employed by individual players, it certainly provides convincing evidence of the value of simulation as a research tool for students of international relations. See Axelrod, *Evolution of Cooperation,* and "An Evolutionary Approach to Norms."

which many fundamental features of the international system were changing simultaneously, it is nearly impossible to separate the impact of the new trade and monetary regimes from the impact of numerous other changes. Those seeking more meaningful cases for use in natural experiments regarding the significance of international regimes would be well advised to consider synchronic comparisons of the various international commodity regimes or the collection of wildlife regimes that have arisen over the last several decades[48] or diachronic assessments of the Antarctic regime or the regimes for pollution control that have emerged in recent years under the tutelage of the regional-seas program of the United Nations Environment Programme (for example, the Mediterranean Action Plan).[49]

Behavioral Models

The argument of the preceding section also suggests the importance of rethinking and refining the theoretical models we rely on to comprehend the behavior of the actors in international relations and, especially, the links between the behavior of individual actors and collective behavior. The analysis of international behavior embedded in the literature on international regimes has struck a remarkably responsive chord among some readers, even while provoking sharply antagonistic reactions among others. In my judgment, these reactions cannot be attributed to substantive results flowing from the study of international regimes so far. Rather, it strikes me that both reactions stem more from linkages between propositions incorporated in the regimes literature and the (often implicit) behavioral models that students of international relations employ. To see the relevance of this observation for any effort to assess the significance of international regimes, consider the following contrasts.

Those who call themselves realists assume, in the final analysis, that the actors in international arenas are status maximizers.[50] Whatever the ultimate source of this behavior, status-maximizing actors evaluate their

48. For a descriptive account of a wide range of wildlife regimes consult Simon Lyster, *International Wildlife Law* (Cambridge: Grotius Publications, 1985).

49. Peter Haas, "Do Regimes Matter? A Study of Evolving Pollution Control Policies for the Mediterranean Sea," paper presented at the annual meetings of the International Studies Association, April 1987.

50. For a brief but clear account of status-maximizing behavior by an economist see E. J. Mishan, *What Political Economy Is All About* (Cambridge: Cambridge University Press, 1982), chap. 17.

own performance in relationship to the performance of others, and they strive to attain the highest possible rank in the hierarchy of members of the international community.[51] The currency for measuring and evaluating performance in this setting is power, a practice that makes sense since power itself is a relational concept.[52] Any gain in power on the part of one actor signifies a relative loss of power for other members of the community and vice versa. It is easy to see that a society composed of status maximizers will resemble a Hobbesian state of nature. Because gains for individual actors amount to losses for others, the members of the group will find themselves in more or less pure conflict (game theorists call it "zero-sum" conflict) with each other all the time. They will be preoccupied with enhancing their own status or avoiding the danger of losing ground to other members of the group; they will not be concerned with the pursuit of feasible joint gains. In such a world, actors may enter into short-term alliances on an expediential basis, and particularly powerful status maximizers may impose their will regarding institutional arrangements on weaker neighbors. But such actors will not experience incentives to accept social contracts in which all members of the group join in restrictive practices or comply with limiting behavioral prescriptions in the interests of realizing joint gains. There simply will not be any feasible joint gains for actors to cooperate in attaining. Analysts who think in these terms typically regard the regimes literature as naive and misguided at best. More than that, they may well treat this literature as dangerous because it diverts attention from the analysis of the dynamics of power, which should be the central concern of students of international relations.

An alternative model borrows heavily from the behavioral assumptions of microeconomics in conceptualizing the actors in international relations as self-interested utility maximizers. This approach, which such writers as Keohane, Krasner, and Stein often espouse, is sometimes labeled (misleadingly in my judgment) neorealism or structural realism.[53] Self-interested utility maximizers choose continuously among

51. Whereas writers like Waltz sometimes suggest that the character of the international system compels individual actors to operate as status maximizers, others attribute such behavior simply to envy, an individual lust for power, or genetic programing of the sort emphasized by the sociobiologists. See Kenneth N. Waltz, *Theory of International Politics* (Reading: Addison-Wesley, 1979).

52. David A. Baldwin, "Money and Power," *Journal of Politics* 33 (1971), 578–614.

53. Keohane engages in some interesting efforts to vary the behavioral assumptions he employs in thinking about international cooperation (*After Hegemony*, 110–132). Nonetheless, the theoretical force of the analysis set forth in *After Hegemony* stems directly from behavioral models that assume that individual actors are self-interested utility maximizers. See also the selection of essays included in Robert O. Keohane, ed., *Neorealism and Its Critics* (New York: Columbia University Press, 1986).

available alternatives in such a way as to maximize their own welfare without comparing their performance with that of others or even considering the implications of their own actions for others. It is easy enough to see why actors of this type will often accept the constraints associated with regimes or other social institutions on the basis of straightforward calculations of a utilitarian nature. To the extent that they find themselves involved in interdependent decision making (a pervasive condition in the realm of international relations), such actors are bound to discover that individually rational behavior regularly produces suboptimal (sometimes even disastrous) outcomes and that they (as well as other members of the group) can benefit from accepting the constraints of institutional arrangements. The significance of institutions in a behavioral environment of this type arises from the facts that individual defectors may find themselves excluded from receiving the benefits produced by regimes and that defection on the part of individuals may precipitate an unraveling of institutional arrangements with a consequent loss of benefits to all members of the group.[54] Furthermore, the transaction costs and collective-action problems entailed in restructuring existing institutional arrangements are sufficient to create an inertial force favoring the status quo. Note, however, that self-interested utility maximizers may be perfectly happy with spontaneous institutional arrangements in contrast to negotiated regimes. What is more, they do not comply with the provisions of prevailing institutions as a result of reinforcement learning (for example, an acquired habit of obedience) or out of any sense of obligation or propriety. They participate in regimes fundamentally as a means of maximizing net benefits for themselves. It follows that those whose behavior emanates from such calculations will feel no compunctions about violating institutional requirements, if and when they conclude that it is possible to increase their net benefits by doing so.[55]

A third behavioral model treats the actors in international relations as occupants of more or less well-defined roles whose actions are heavily constrained by the requirements of the roles they occupy. Though Krasner labels this perspective Grotian,[56] it is actually easier to comprehend the essential character of the model in terms of Rawls's analysis of practices. As Rawls puts it, "It is the mark of a practice that being taught

54. For some related observations see Robert O. Keohane, "Reciprocity in International Relations," *International Organization* 40 (1986), 1–27.
55. To the extent that such actors are rule utilitarians in contrast to act utilitarians, however, it is reasonable to assume that they will not continuously contemplate the pros and cons of violating the detailed requirements of specific regimes.
56. Krasner, "Structural Causes and Regime Consequences," 8–10.

how to engage in it involves being instructed in the rules that define it, and that appeal is made to those rules to correct the behavior of those engaged in it. Those engaged in a practice recognize the rules as defining it. The rules cannot be taken as simply describing how those engaged in the practice in fact behave; it is not simply that they act as if they were obeying the rules."[57] Participation in a practice, "necessarily involves the abdication of full liberty to act on utilitarian or prudential grounds."[58] On this account, actors may have an opportunity to choose which practices to join; it is possible to opt for football rather than baseball or for nonalignment rather than alignment at the international level. More often, however, individuals have little choice about the roles they occupy, because they are socialized into accepting certain roles without question or because they acquire their roles through an interactive learning process that does not involve conscious choice.[59] But in either case, individual actors are not at liberty to violate the rules or conventions of a practice once they have become participants in it. It just does not make sense for a chess player to refuse to accept the concept of checkmate, for a speaker of English to assert that it makes no difference whether subjects and predicates agree, or for an actor in the existing international society to disregard the rules governing the nationality of citizens or the use of the electromagnetic spectrum. In the international community, the menu of practices available to individual parties is almost always sharply limited. A "new" state, for example, has little choice but to join the basic institutional arrangements of the states system. As renegade states like the Soviet Union in the 1920s and the Peoples Republic of China in the 1960s discovered, there really are no other practices available at this level. With respect to more specific regimes, there is sometimes more latitude for choice. But even in these areas the limitations are severe. What are the options, for example, for a non-socialist state that dislikes the provisions of the prevailing trade and monetary regimes or for any state that objects to the provisions of the existing regime for the electromagnetic spectrum or emerging coastal-state regimes for the marine fisheries?

To return now to the basic issue, it is apparent that these behavioral models have radically different implications regarding the significance of international regimes. There is little or no room for social institutions (other than highly transient or purely coercive arrangements) in a world of status maximizers. Regimes may well become critical determinants of

---

57. John Rawls, "Two Concepts of Rules," *Philosophical Review* 65 (1955), 24.
58. Ibid.
59. For an account of such processes see Axelrod, "An Evolutionary Approach to Norms."

collective behavior in a world of self-interested utility maximizers. Yet institutional arrangements can also run down or unravel with remarkable speed in such a behavioral environment. By contrast, practices or institutional arrangements may come to dominate a world of role players.

It is therefore of great importance to weigh the relative merits of these models in thinking about the future of the research program embedded in the regimes literature. My own view is that there is no need to make a definitive choice among these behavioral models. Like competing models in other fields of enquiry (for example, the wave and particle theories of light), each captures some important features of reality but none offers a satisfactory account of the full range of observable phenomena. Even so, many students of international relations, especially those who style themselves realists, will want to argue that the behavior of international actors is weighted more toward status maximizing than the typical behavior of actors in domestic settings. But this proposition is surely difficult to sustain in a convincing manner. Partly, this is because status-maximizing behavior is widespread in domestic settings. There is nothing uncommon in domestic society about individuals motivated by a desire to keep up with the Joneses or, more simply, by envy. In part, this conclusion arises from the fact that both utility-maximizing behavior and role-playing behavior are common at the international level. The observation that much of this behavior occurs in connection with social institutions that do not feature highly developed organizations should not be allowed to cloud our vision.[60] In any case, I am persuaded that our conclusions regarding the significance of institutional arrangements as determinants of collective behavior at the international level will ultimately rest heavily on the judgments we make about the status of these alternative behavioral models. If we want to reach definite conclusions about the validity of the vision of the world embedded in the literature on regimes, therefore, we must examine our assumptions about the behavior of the actors in international arenas more carefully.

THE PROMISE OF INSTITUTIONALISM

The preceding discussion makes clear why the jury is still out on the fundamental question of whether the recent surge of interest in international regimes represents a development of lasting importance, rather than a mere fad or fashion, in our thinking about international relations.

60. See also Keohane, "Reciprocity in International Relations."

To answer this question we must come to terms with the issue of whether social institutions are major determinants of collective outcomes at the international level rather than simply continue to explore the conditions governing stability and change in institutional arrangements. Because the investigation of social institutions as determinants of collective international outcomes involves our most basic assumptions about the behavior of actors in international arenas, we will only be able to provide a conclusive answer to this question by rethinking and refining some of the major premises on which the study of international relations rests.

Even so, the literature on international regimes has already played a constructive role in promoting a reintegration of the subfields of international politics, economics, law, and organization. It is no accident that this literature is closely associated with the ideas of those who have become leaders in the movement to revive the study of international political economy. Whether we look at the international-trade regime, the international monetary regime, and the international oil regime,[61] or focus on the institutions governing marine fishing, deep-seabed mining, maritime commerce, and the treatment of radioactive fallout, it is apparent that these arrangements involve a complex interaction of politics and economics in which the actions of states and private entities impinge on each other at every turn. Any effort to comprehend the politics or the economics of these regimes in isolation would be severely limited, a fact that makes the choice of a political economy perspective in thinking about such arrangements entirely natural.

It would also be difficult to achieve a sophisticated grasp of the character of these regimes without thinking systematically about questions of international law and organization. More often than not, the institutional arrangements included in these regimes are embodied in conventions or treaties that require careful interpretation and adaptation to changing circumstances, just as municipal legal formulas do. Similarly, the relationships between the regimes themselves and various explicit organizations is a subject of obvious importance in this field of enquiry. Exactly how critical is the presence of an organization like the International Monetary Fund to the operation of the prevailing international monetary regime? Why are organizations seemingly less important in connection with the international-trade regime than in connection with the international monetary regime? To what extent is the heavy reliance placed on the International Seabed Authority in the regime for deep-seabed mining likely to become a liability? How important is it to set up a separate organization to handle problems of reconstruction and com-

61. Keohane, *After Hegemony*, 135–181.

pensation in order to avoid undermining the role of the International Atomic Energy Authority in conjunction with the emerging regime for nuclear accidents? Most of those who seek to understand international regimes, therefore, have become students of international relations in the true sense of the term, endeavoring to shed light on the interfaces among international politics, economics, law, and organization. Whatever the ultimate fate of the literature on international regimes, this feature of it is surely an extremely valuable one that we should do our best to preserve and nurture.

CHAPTER NINE

Praxis: Institutional Design in International Society

Policymakers and students of international affairs alike commonly focus on the operation of institutional arrangements or international regimes in their efforts to explain collective outcomes occurring in international society. We are all familiar, for instance, with the argument that wars are stimulated or even caused by the anarchical character of the states system. This line of reasoning has informed the thinking of a wide range of reform groups (for instance, world federalists and those promoting world peace through world law) seeking to restructure the institutions prevailing in international society.[1] It has also enjoyed considerable currency in the academic community, at least over the last several generations.[2] More specifically, the extraordinary growth of international trade during the postwar era is commonly attributed to the establishment of a liberalized regime for international trade. Many observers have blamed the international monetary crises of the 1960s and early 1970s on the system of fixed exchange rates operative at the time or on other features of the postwar international monetary regime. Severe depletions of renewable natural resources, such as fish stocks in marine areas, are widely interpreted as consequences of regimes featuring relatively unrestricted common-property arrangements. Severe malnutrition in various parts of the Third World is frequently regarded as an outcome of the international food regime, which relies on competi-

---

1. For illustrations see Cord Meyer, *Peace or Anarchy* (Boston: Little, Brown, 1947); Emory Reves, *The Anatomy of Peace* (New York: Harper, 1945); and Grenville Clark and Louis B. Sohn, *World Peace through World Law* (Cambridge: Harvard University Press, 1958).

2. Such views constitute what Waltz calls third-image analyses. See Kenneth N. Waltz, *Man, the State, and War* (New York: Columbia University Press, 1959).

tive markets as the basic mechanism for distributing food at the international level.[3] Even the difficulties afflicting efforts to create an effective transnational communications network by stationing satellites in geosynchronous orbits are commonly linked to the character of the prevailing regime governing international broadcasting.[4]

Those who adopt this point of view are naturally drawn to social engineering, or, as it is now fashionable to call it, institutional design at the international level.[5] Emphasizing institutional design, which directs attention to deliberate or planned alterations in institutional arrangements in the interests of achieving identifiable goals, lends policy relevance to the analysis of international regimes and offers hope for solving problems of the sort referred to in the preceding paragraph. Approached from this perspective, propositions dealing with developmental sequences and patterns of change in international regimes not only deepen our understanding of regime dynamics, they also add to our ability to institute planned changes in institutional arrangements. More broadly, an emphasis on institutional design offers human communities a basis for believing that they can exert some control over their own destinies. A search for understanding of the sources and patterns of change in international regimes is certainly interesting as an end in itself. But it is not sufficient to permit us to escape the bonds of social determinism.[6] To the extent that we acquire the ability to design and implement institutional arrangements such as international regimes, however, it is possible to attain a meaningful sense of fate control.

What is more, the twentieth century has witnessed a number of striking efforts to design and implement international institutions on a deliberate basis. The creation of the League of Nations and, subsequently, the establishment of the United Nations are properly understood as exercises in institutional design. And there have been many attempts to design international regimes of a more functionally specific nature. A

3. See also·A. K. Sen, *Poverty and Famines: An Essay on Entitlement and Deprivation* (Oxford: Clarendon Press, 1981).

4. For a selection of analyses dealing with these functionally specific regimes see the applied essays in Stephen D. Krasner, ed., *International Regimes* (Ithaca: Cornell University Press, 1983). On the case of international broadcasting consult Seyom Brown, Nina W. Cornell, Larry L. Fabian, and Edith Brown Weiss, *Regimes for the Ocean, Outer Space, and Weather* (Washington, D.C.: Brookings, 1977), and Gregory C. Staple, "The New World Satellite Order: A Report from Geneva," *American Journal of International Law* 80 (1986), 699–720.

5. For a more general discussion of institutional design treated as applied public choice theory see Clifford S. Russell, ed., *Collective Decision Making: Applications from Public Choice Theory* (Baltimore: Johns Hopkins University Press, 1979).

6. For a more general account of social determinism consult Ernest Nagel, *The Structure of Science* (New York: Harcourt, Brace and World, 1961), 592–602.

regime to govern commercial whaling was created in the 1930s and reestablished under the terms of the international whaling convention of 1946. The Bretton Woods agreements of 1944 set up a relatively elaborate international monetary regime, which played a role of some importance in reconstructing the international economic order during the first postwar generation. Various members of international society have created specific regimes to stabilize international transactions in commodities such as tin, sugar, coffee, and rubber. The 1959 Treaty on Antarctica lays out a regime to regulate human activities on the Antarctic continent. Similarly, the Third United Nations Conference on the Law of the Sea provided an arena for extensive negotiations in the 1970s and early 1980s aimed at reaching agreement on a comprehensive system of institutional arrangements for marine areas and resources.

Yet we need only consider a few examples to realize that institutional design, in the sense of conscious efforts to devise properly functioning international regimes, is a tricky business. Opportunities to create regimes that would prove beneficial to all participants are regularly lost. The field of arms control is replete with dramatic examples of this phenomenon, but the history of negotiations concerning a variety of transboundary environmental issues makes it clear that it occurs in other functional areas as well. Even after they are put in place, international regimes often fail to yield the intended results. Most of the commodity arrangements, for example, have performed indifferently, leaving significant loopholes that have been exploited by those desiring to manipulate these arrangements or collapsing under the pressure of shifting political circumstances.[7] What is more, regimes frequently produce outcomes unforeseen and unintended by their creators. To illustrate, the international whaling regime, established to regulate commercial whaling, has become an increasingly effective vehicle for those determined to put an end to commercial whaling altogether. And it seems reasonable to conclude that the Antarctic regime has stimulated the growth of interest in exploiting the natural resources of Antarctica and the Southern Ocean by promoting an atmosphere of political stability and providing a stimulus for extensive exploratory and research efforts in the region. Beyond this, international regimes are regularly overtaken by circumstances that were not anticipated by those who designed them. The international monetary regime has experienced dramatic shocks widely attributed to the erosion of the dominance of the United States in the international economic order. Regimes designed to ensure max-

7. Mark W. Zacher, "Trade Gaps, Analytical Gaps: Regime Analysis and International Commodity Trade Regulation," *International Organization* 41 (1987), 173–202.

imum sustained yield from marine-mammal populations, such as the arrangement for northern fur seals, have been overtaken by a surging tide of opposition to all consumptive uses of marine mammals.

How can we account for the resultant combination of intense interest in the design of institutional arrangements at the international level coupled with pronounced limitations on our ability to engage successfully in social engineering in this realm? Are there significant opportunities to improve our capacity to design international regimes at a reasonable cost? The next two sections of this chapter contain an analysis of the constraints facing those interested in institutional design in international society. This analysis sets the stage for the subsequent discussion of a design strategy tailored to enhance the prospects for success in future efforts to establish international regimes on a deliberate basis.[8]

INFORMATION CONSTRAINTS

True hegemons, acknowledged leaders of privileged groups, or Arrowian dictators are few and far between in international society.[9] These concepts point to extreme (if not ideal) types, which provide benchmarks for the evaluation of real-world situations rather than descriptions of actual cases. Even so, it is useful to begin this discussion with these extreme types because doing so allows us to abstract away the limitations on institutional design in international society attributable to the dynamics of collective action and, therefore, to focus on some impediments to the design of international regimes that have nothing to do with collective action. A brief account should suffice to demonstrate that these impediments are far from trivial.

Social institutions, including international regimes, are complex and poorly understood phenomena. Several factors limit efforts to model international regimes in a rigorous fashion. Families of institutional arrangements, such as regimes featuring floating exchange rates for monetary transactions or limited-entry mechanisms for renewable re-

8. For a similar discussion directed toward institutional arrangements in domestic societies see Geoffrey Brennan and James M. Buchanan, *The Reason of Rules: Constitutional Political Economy* (Cambridge: Cambridge University Press, 1985), chap. 9.

9. A privileged group is one containing a single member who values some public good highly enough to supply the good regardless of the behavior of other group members. For a seminal account see Mancur Olson, Jr., *The Logic of Collective Action* (Cambridge: Harvard University Press, 1965). An Arrowian dictator exists under conditions such that "whenever the dictator prefers x to y, so does society" (Kenneth J. Arrow, *Social Choice and Individual Values,* 2d ed. [New York: Wiley, 1963], 30).

sources, commonly include numerous variants differing significantly in the collective outcomes they produce. Even if we assume that individual actors are rational and self-interested in some straightforward sense of these notions, therefore, it is not easy to pin down the incentives associated with specific institutional arrangements. Additionally, the well-known problems of extrapolating from micromotives to macrobehavior complicate efforts to evaluate alternative regimes in analytic terms.[10] We are, after all, finally interested in collective outcomes, such as the stability of currencies in relationship to each other or the viability of fish stocks, in contrast to the incentives facing individual actors in designing international regimes. Additionally, there are compelling reasons to expect the outcomes flowing from specific institutional arrangements to be affected significantly by the nature of the broader social setting and the character of the larger institutional arrangements or orders within which they are nested. Though relatively unrestricted common-property arrangements worked reasonably well for high-seas fisheries in an era of light usage, for instance, they collapsed in short order following the introduction of high-endurance stern trawlers.[11] Similarly, international monetary arrangements featuring fixed exchange rates, which worked perfectly well in earlier eras, cannot survive in a period in which the pursuit of domestic prosperity and full employment takes precedence over maintaining the strength of national currencies in international transactions.

Faced with these analytic constraints, it is certainly possible to proceed empirically, examining outcomes that have flowed from real-world international regimes in the hope of developing useful empirical generalizations about institutional arrangements in international society. Yet this is easier said than done when it comes to the study of complex social institutions. Not only are there self-evident limits on the feasibility of conducting true experiments with institutional arrangements, it is also tricky to devise worthwhile natural or field experiments relating to international regimes. The universe of comparable cases tends to be small in this context. In fact, superficially similar cases can easily turn out to differ in ways that cast doubt on the validity of including them in the same group of cases for purposes of empirical analysis. Moreover, variations in social settings make it difficult to formulate generalizations that hold up well across space and time.[12] As Goldberg puts it, therefore,

10. Thomas C. Schelling, *Micromotives and Macrobehavior* (New York: Norton, 1978).

11. For a dramatic account see William Warner, *Distant Water* (Boston: Little, Brown, 1983).

12. For an account that makes it clear that this problem is widespread in the social sciences see Nagel, *Structure of Science*, 459–466.

"the results stemming from the establishment of new institutions or modifications in existing ones are seldom known precisely and are often widely divergent from the original expectations."[13]

Problems relating to value trade-offs also plague many efforts to design institutional arrangements. Even if the participants make use of broad or encompassing formulations of utilitarian values, such as allocative efficiency, it is to be expected that those involved in the design of institutional arrangements at the international level will want to maximize the attainment of several distinct values simultaneously.[14] Over and above efficiency, these values are likely to include equity, cultural diversity, and ecological integrity. It is, of course, conceivable that some specific institutional arrangement will constitute a dominant solution in a given issue area if it turns out to be preferable to every alternative with respect to some value(s) and at least as good as all the alternatives with respect to every value. But this is a remote prospect under real-world conditions. In the usual cases, different regimes will seem preferable in terms of one or another of the value criteria.

To make the resultant problem more concrete, consider the task of designing an optimal international regime to govern the commercial exploitation of deep-seabed minerals. Decision theory offers an array of more or less sophisticated techniques for solving problems of this kind.[15] But these techniques are likely to prove no more than heuristically useful in coming to terms with the case at hand. How can we construct a transformation function characterizing the set of feasible trade-offs between efficiency in the extraction of minerals from the deep seabed and equity with regard to access to the minerals as well as the distribution of the proceeds when we are not quite sure how to measure either of these values, much less how to determine the rate at which we can obtain one of them at the expense of the other (that is, the marginal rate of substitution between efficiency and equity)?[16] And these problems are relatively straightforward by comparison with the difficulties associated with ef-

13. Victor P. Goldberg, "Institutional Change and the Quasi-Invisible Hand," *Journal of Law and Economics* 17 (1974), 482. Note, however, that imperfect information may actually ameliorate bargaining processes involved in the formation of some regimes by making the participants less dogmatic about their preferences. See, for example, Howard Raiffa, *The Art and Science of Negotiation* (Cambridge: Harvard University Press, 1982), 273–274.

14. See also the essays in Laurence Tribe, Corinne S. Schelling, and John Voss, eds., *When Values Conflict* (Cambridge: Ballinger, 1976).

15. Consult, among others, Ralph Keeney and Howard Raiffa, *Decisions with Multiple Objectives: Preferences and Value Tradeoffs* (New York: Wiley, 1976).

16. Okun, for example, describes the trade-off between efficiency and equity as "the big tradeoff." See Arthur M. Okun, *Equality and Efficiency: The Big Tradeoff* (Washington, D.C.: Brookings, 1975).

forts to construct indifference maps pertaining to such value trade-offs. How much would we be willing to sacrifice in terms of efficiency (measured, let us say, in terms of the contribution of deep-seabed mining to gross world product) in order to secure an international seabed regime producing results that conform to our standard of equity (conceptualized, for instance, in terms of some notion of equality of opportunity for participants in the mining activities or some meausre of justice in the distribution of the economic returns to the minerals themselves)? The average member of international society does not possess fully articulated preferences regarding issues of this sort; some have not addressed such questions in a meaningful fashion at all. Without denying the existence of increasingly sophisticated techniques designed to elicit revealed preferences, therefore, it seems doubtful that directly interpretable indifference maps for the trade-offs under consideration here can be constructed.[17]

As well, the task under consideration here would require some procedure for aggregating the indifference maps of the individual members of international society to arrive at a social-welfare function pertaining to an international seabed regime. Yet no satisfactory technique exists for identifying social-welfare functions regarding comparatively simple choices among conventional goods and services, much less one that would suffice to identify community preferences with respect to complex institutional arrangements. It would therefore be hard to make unambiguous recommendations about the relative desirability of alternative regimes to govern deep-seabed mining (or any other similar activity), even if we were able to predict the actual results that would flow from the selection of specific alternatives with reasonable accuracy. In sum, as Hitch has observed, existing "tools are far from satisfactory for predicting and assessing the effects of suggested institutional changes."[18]

INSTITUTIONAL DESIGN AND COLLECTIVE ACTION

As I have suggested throughout this book, moreover, regime formation in international society ordinarily involves some form of collective action. This is not to say that those participating in the development of specific regimes are equal with respect to the effective power, influence, or bargaining strength they can bring to bear on the selection of institu-

17. For a review of modern techniques for preference revelation see Dennis Mueller, *Public Choice* (Cambridge: Cambridge University Press, 1979), chap. 4.
18. Charles J. Hitch, "Introductory Remarks," in Russell, *Collective Decision Making*, xiii.

tional arrangements at the international level. Far from it. Even so, situations in which one or a few members of international society are able to dominate the formation of regimes, as in the development of the international monetary regime in the 1940s, are the exception rather than the rule. In the typical case, institutional arrangements emerge from more or less complex interactions among a multiplicity of actors. It follows that all those interested in institutional design in international society must devote considerable attention to the nature of collective action and the implications of collective action for planned alterations in institutional arrangements.

Some interactions involving the form or content of international regimes are either distributive, as opposed to allocative, in character or perceived as distributive by the parties engaged in the interaction.[19] To the extent that such perceptions prevail, negotiations regarding the creation or adaptation of regimes will differ markedly from the contractarian encounters envisioned by analysts such as Rawls and Buchanan.[20] Those endeavoring to devise a social contract are generally seen as starting with a blank slate (for example, the status quo in Rawls's original position)[21] or with a preexisting situation that is highly unsatisfactory to all concerned (for example, a Hobbesian state of nature or a severe crisis).[22] Their task is to create institutional arrangements designed to promote the common good over the long run. Under such conditions, it is easy to imagine a variety of institutional arrangements that would yield gains for all parties concerned or that would, in other words, be Pareto superior to the status quo. By contrast, actual efforts to adjust international regimes often involve alterations in prior bargains or preexisting social contracts. In such cases, those benefitting from the preexisting arrangements will naturally suspect that the proposed adjustments will improve the outcomes for others at their expense. Whenever this occurs, suggested changes in institutional arrangements may come to be treated as redistributive issues rather than occasions for innovations likely to produce mutual gains. There is, of course, nothing peculiar about international society in these terms; efforts to promote institu-

19. The distinction between distributive and allocative concerns dates at least from the work of Wicksell. On the modern literature pertaining to this distinction see Mueller, *Public Choice*, chap. 14.

20. John Rawls, *A Theory of Justice* (Cambridge: Harvard University Press, 1971), and James Buchanan, *The Limits of Liberty* (Chicago: University of Chicago Press, 1975).

21. For Rawls's own account of the character of the original position see ibid., 17–22.

22. Note, however, that the Hobbesian vision of the state of nature as an unpleasant and dangerous environment in which life is apt to be nasty, brutish, and short is not shared by all political philosophers. Rousseau, for example, envisioned the state of nature as a rather attractive setting only gradually corrupted by the acquisitive activities of humans.

tional reform are often fueled by redistributive motives in all social systems. Yet issues of this type are invariably difficult to handle in a nondisruptive fashion; they are apt to trigger competitive forays into coercive diplomacy whenever the participants possess intense or passionate feelings about them.[23] Under the circumstances, the processes involved in settling redistributive issues seldom conform well to the vision of planning and reasoned choice embedded in the concept of institutional design.

These observations go far toward explaining the striking differences between the negotiations leading to the Bretton Woods agreement on a new monetary regime in 1944 or to the provisions of the Antarctic Treaty of 1959 on the one hand and the negotiations associated with the articulation of the Smithsonian agreements of 1971, the bargaining involved in hammering out a regime to govern deep-seabed mining, and efforts to introduce a new international economic order on the other. The preexisting international monetary regime had effectively collapsed under the onslaughts of the Great Depression and World War II, and there never had been an international regime applicable to Antarctica. Accordingly, the parties negotiating the Bretton Woods agreements and the Antarctic Treaty System found it comparatively easy to think it contractarian terms; the creation of a new regime offered joint gains for all those involved in each case. Contrast these circumstances with those arising in connection with the negotiations regarding new regimes for marine resources and, especially, the debates relating to a new international economic order. Recent trends in the regimes for marine resources involve alterations that are regarded by many as detrimental to the interests of distant-water states, certain industrialized states, and states desiring maximum freedom to carry out scientific research. Even more to the point, proposals for an array of institutional changes grouped under the rubric of a new international economic order are typically interpreted by the advanced industrialized states as initiatives on the part of the less developed states intended to alter established institutional arrangements in their favor.

Even when participants can envision a zone of agreement or a contract zone with regard to changes in some existing international regime, efforts to reach agreement on a specific point within this zone frequently give rise to hard bargaining in contrast to the reasoned deliberations

23. For a well-known account of coercive diplomacy with particular reference to international society see Thomas C. Schelling, *Arms and Influence* (New Haven: Yale University Press, 1966). See also Daniel Ellsberg, "The Theory and Practice of Blackmail," 343–363 in Oran R. Young, ed. and contributor, *Bargaining: Formal Theories of Negotiation* (Urbana: University of Illinois Press, 1975).

envisioned by many of those seeking to promote institutional design or social engineering at the international level.[24] To see the implications of this proposition, contrast the situation of Rawls's individuals endeavoring to select a set of institutional arrangements behind the "veil of ignorance"[25] with the situation of real-world actors bargaining over concrete proposals for alterations in some specific international regime.[26] Because no one possesses specific knowledge of his role in society and, therefore, of his actual interests behind the veil of ignorance, each participant in Rawls's original position is motivated to concentrate on designing institutional arrangements that will serve to maximize social welfare or to promote the common good.[27] In one stroke, the use of this device transforms negotiations regarding social institutions into exercises in institutional design, though it does not eliminate the technical problems associated with such exercises (as the protracted controversy over Rawls's specific proposals makes clear).[28] By contrast, parties to real-world interactions concerning adjustments to preexisting international regimes often know perfectly well who they are and what the interests of their states are. In such cases, individual participants will evaluate all proposals in the light of their implications for the interests of their own states and bargain vigorously to achieve agreements yielding the best possible outcomes for their states. It should come as no surprise, under the circumstances, that efforts to agree on the terms of an international seabed regime proved contentious or that parties sometimes find it hard to agree on institutional arrangements even in such relatively technical areas as the use of the electromagnetic spectrum.

24. On the concept of a zone of agreement see Raiffa, *Art and Science of Negotiation*. Note also that certain subnational interest groups may anticipate losses in connection with proposed alterations in institutional arrangements, even though the alterations can be expected to yield net benefits for all the relevant states in the aggregate. To the extent that such subnational groups are able to veto or block actions on the part of states, they may succeed in preventing changes in prevailing regimes despite the existence of a zone of agreement.

25. The veil of ignorance is a hypothetical construct that Rawls and others employ in their efforts to think about optimal institutional arrangements for societies. For Rawls's own account of the veil of ignorance see Rawls, *A Theory of Justice*, 136–142.

26. For a richly descriptive case in point see I. William Zartman, *The Politics of Trade Negotiations between Africa and the EEC* (Princeton: Princeton University Press, 1971).

27. Buchanan attempts to achieve similar results by focusing on the veil of uncertainty under which parties are induced to think in terms of the common good because they cannot predict how complex institutional arrangements will affect their individual welfare over the long run. See, for example, Brennan and Buchanan, *Reason for Rules*, 28–31.

28. That is, there is considerable controversy regarding the sorts of social institutions that those negotiating a social contract from the original position behind a veil of ignorance would ultimately adopt. See, for example, Brian Barry, *The Liberal Theory of Justice* (Oxford: Oxford University Press, 1973), and Robert Paul Wolff, *Understanding Rawls*, (Princeton: Princeton University Press, 1977).

Observe also that bargaining over the terms of international regimes commonly takes the form of interactions involving many parties and many issues.[29] Though there is certainly scope for structuring negotiations that deal with institutional arrangements in such a way as to enhance the prospects of arriving at generally desirable outcomes, it is virtually never feasible to reduce interactions pertaining to complex international regimes to two parties and a single issue. The complex bargaining that results may open up possibilities for building consensus through logrolling and efforts to devise package deals. But such bargaining is also fraught with additional complications that pose problems for those desiring to engage in institutional design with regard to international regimes. Much of the energy that goes into multiparty negotiations is often consumed in efforts to form and reform effective coalitions rather than to evaluate the relative merits of alternative institutional arrangements as such.[30] Despite the existence of a distinct zone of agreement, the parties may well fail to reach agreement on any departure from the status quo as a result of strategic misrepresentations in the transmission of information about their preferences or the deployment of bargaining tactics leading to stalemates.[31] As all students of negotiation know, such failures to realize perfectly feasible joint gains are common occurrences in bargaining situations. Moreover, agreements that do emerge from such interactions often take the form of incoherent compromises or loosely textured formulas adopted to avoid serious efforts to reconcile major conflicts of interest. Agreements of this sort simply shift many of the real problems of devising institutional arrangements to the implementation phase, and they may also sow the seeds of additional, unintended changes. Certainly, we can hope that the process of implementation will prove more amenable to those endeavoring to engage in social engineering at the international level.[32] But it is difficult to equate the negotiated transitions that have produced many of the institutional arrangements operative in international society with the results of successful ventures in institutional design.[33]

29. For an extensive account of the significance of these factors consult Raiffa, *Art and Science of Negotiation,* and James K. Sebenius, "Negotiation Arithmetic: Adding and Subtracting Issues and Parties," *International Organization* 37 (1983), 281–316.

30. The analysis of the resultant processes of coalition formation constitutes the central concern of *N*-person game theory. For a standard treatment consult R. Duncan Luce and Howard Raiffa, *Games and Decisions* (New York: Wiley, 1957).

31. Thomas C. Schelling, *The Strategy of Conflict* (Cambridge: Harvard University Press, 1960).

32. More often than not, however, interested parties simply continue their efforts to gain advantages within those arenas charged with responsibility for implementation. On the resultant bureaucratic politics see Graham Allison, *The Essence of Decision* (Boston: Little, Brown, 1971).

33. Even when such transitions do yield coherent results, moreover, the transaction

These observations go far toward explaining the difficulties that plagued the recent law-of-the-sea negotiations as well as many negotiations pertaining to the development of environmental regimes at the international leve.[34] Given the opportunities for linking issues and arriving at package deals through logrolling, it seems reasonable to conclude that there was room during the 1970s and 1980s for the creation of a set of institutional arrangements for marine areas and resources yielding net benefits for all participating states.[35] Yet the members of international society experienced severe difficulties in reaching a generally satisfactory agreement. In the end, many institutional arrangements for the oceans may well assume the form of spontaneous or self-generating regimes instead of negotiated regimes. There is no doubt that the negotiations of the 1970s and early 1980s regarding institutional arrangements for the oceans were affected by several of the problems referred to in the preceding paragraph, such as coalitional politics and rigidities arising from the efforts of some participants to make use of bargaining tactics. Prospects for reaching a generally acceptable negotiated settlement suffered in this case, moreover, from shifts in the preferences (or tastes) of key participants during the course of the negotiations. This is surely the proper interpretation, for instance, of the sharp shift that took place in the posture of the United States in these negotiations during 1981.

Given these features of collective-action processes, it should come as no surprise that some commentators have sought to interpret various limitations on institutional design with regard to international regimes as virtues. Those who approach institutional change as a spontaneous or self-generating process, for example, often equate institutional change with the interactions taking place in the context of competitive markets; they expect some invisible hand to yield socially desirable adjustments in international regimes as needed.[36] While this perspective does not generate the sense of efficacy derivable from a faith in institutional design, it

---

costs are apt to be high, thus adding yet another complication to the situation facing those struggling to design institutional arrangements properly adapted to the prevailing international environment.

34. For a rich descriptive account focusing on efforts to devise arrangements to control marine pollution see R. Michael M'Gonigle and Mark Zacher, *Pollution, Politics, and International Law* (Berkeley: University of California Press, 1979).

35. See Raiffa, *Art and Science of Negotiation*, chap. 18, and Sebenius, "Negotiation Arithmetic."

36. Some commentators, for example, envision the operation of an invisible hand guiding the evolution of common or customary law. See Friedrich A. Hayek, *Rules and Order*, vol. 1 of *Law, Legislation, and Liberty* (Chicago: University of Chicago Press, 1973), and Anthony Scott, "Property Rights and Property Wrongs," lecture delivered to the Canadian Economics Association, June 1983.

does lead to the comforting conclusion that there is little need to worry about the problems of designing suitable international institutions on a conscious basis. Alternatively, some observers have sought to make a virtue out of the imposition of institutional arrangements by hegemonic actors or ruling coalitions. Focusing on the proposition that international regimes exhibit the attributes of public goods to a relatively high degree, they look upon the membership of such regimes as Olsonian privileged groups in which the consequences of free riding are fortuitously ameliorated, if not eliminated altogether.[37] Here, too, the limitations on institutional design at the international level may not seem particularly troubling.

Yet each of these lines of reasoning is ultimately unconvincing. Though spontaneous or self-generated transitions do occur with some frequency, there is little basis for assuming that such processes of change will typically yield collective outcomes that are socially desirable. The analogy between spontaneous transitions in complex institutional arrangements and market interactions is heuristic at best. In any case, there are good reasons to expect that the resultant interactions will be heavily affected by functional equivalents of market failures.[38] For its part, the idea that international society or substantial segments of it may take on the characteristics of privileged groups when it comes to the development of institutional arrangements ignores wide variations in valuations of the relevant goods among the members of international society. Not only is it probable that some members of international society will derive far fewer benefits than others from particular regimes, it is also likely that certain states will view any given regime as a public bad rather than as a public good.[39] They may reach such conclusions either because they are forced to participate in a regime against their will (as in the case of certain security regimes) or because they are nonmembers who regard a regime as a discriminatory club (as in the case of some nonmembers of the Antarctic regime or of various commodity regimes). In either case, it would be more accurate to speak of institutionalized exploitation under conditions of this sort than to characterize some set of states as a privileged group.[40]

37. See Mancur Olson, Jr., and Richard Zeckhauser, "The Economic Theory of Alliances," *Review of Economics and Statistics* 48 (1966), 266–279.

38. For a wide-ranging analysis of such matters consult Schelling, *Micromotives and Macrobehavior*.

39. Russell Hardin, *Collective Action* (Baltimore: Johns Hopkins University Press, 1982), 61–66.

40. See also the analysis of such phenomena articulated in world-system terms in Bruce Andrews, "The Political Economy of World Capitalism: Theory and Practice," *International Organization* 36 (1982), 135–163.

A DESIGN STRATEGY FOR INTERNATIONAL SOCIETY

Economists and others schooled in the use of utilitarian models generally assume that rational utility maximizers will reach agreement on
mutually beneficial institutional arrangements, including international
regimes, whenever a contract zone or zone of agreement exists. On this
account, the substantive provisions of institutional arrangements flow
from the interactions of interested parties, and it is reasonable to expect
that the outcomes will conform in a general way to the requirements of
Pareto optimality.[41] Self-conscious efforts to design institutional arrangements are therefore largely unnecessary; they may even lead to
interventionist initiatives that contribute to the occurrence of suboptimal results. For their part, political scientists and others who look to the
distribution of power in society as the key to understanding collective
outcomes typically assume that institutional arrangements, such as international regimes, reflect the preferences or interests of the dominant
members of social systems. Since dominant actors ordinarily have a
pretty clear idea of what they want in the way of institutional arrangements, problems of institutional design do not loom large from this
perspective. Rather, the emphasis shifts to an examination of the techniques available to dominant actors as they bring pressure to bear on
others to accept the institutional arrangements they prefer.[42]

The analysis presented in this book suggests that each of these visions
of regime formation in international society is wide of the mark. International regimes typically emerge from complex bargaining processes.
There may be marked asymmetries among the participants in such
processes in terms of their bargaining strength.[43] The relevant interactions may feature explicit bargaining, tacit bargaining, or some mix of
the two.[44] And there may be considerable scope for interested parties to
structure the character of such bargaining processes before well-defined
negotiations get underway. But the important thing to recognize is that

41. Scott, "Property Rights and Property Wrongs."
42. See Arthur A. Stein, "The Hegemon's Dilemma: Great Britain, the United States,
and the International Economic Order," *International Organization* 38 (1984), 355–386;
Stephen D. Krasner, *Structural Conflict: The Third World against Global Liberalism* (Berkeley:
University of California Press, 1985); and Robert Gilpin, *The Political Economy of International Relations* (Princeton: Princeton University Press, 1987).
43. Bargaining strength is a function of a party's ability to hold out for the terms it
prefers in the course of a negotiation. In general, the lower the costs of no agreement to a
party, the greater its bargaining strength. It follows that parties regarded as great powers
in terms of conventional measures of strength will not necessarily possess superior bargaining strength in specific negotiations.
44. Schelling, *Strategy of Conflict,* and Robert Axelrod, *Evolution of Cooperation* (New
York: Basic Books, 1984).

bargaining over the shape or content of international regimes is neither the somewhat mechanical process envisioned by the utilitarians nor the one-sided process envisioned by those who focus on the role of dominant actors. What is more, complex bargaining of this type often fails to result in agreements on institutional arrangements, even when an acknowledged contract zone or zone of agreement exists.[45] As a result, there is considerable scope, in the typical case, for exercising leadership by moving negotiations toward productive outcomes with regard to the choice of institutional arrangements.

Approached from this perspective, institutional design emerges as a process of steering complex bargaining toward coherent and socially desirable outcomes. Such activities diverge, sometimes dramatically, from the sort of rational planning we often associate with the concept of institutional design or social engineering. They place a premium on the ability to guide collective action processes in contrast to any notion of a mechanical application of theoretically grounded propositions about international regimes to concrete situations. But this certainly does nothing to diminish the importance of institutional design in connection with the development of appropriate regimes in international society. What is more, a careful analysis of the complex bargaining associated with regime formation yields a number of lessons for those interested in the design of international institutions. In the following paragraphs, I endeavor to capture the essence of some of the most important of these lessons.

Seize Opportunities

Bargaining over the substantive content of international regimes frequently bogs down as the reciprocal efforts of various parties to make use of bargaining tactics to secure advantages for themselves lead only to stalemate. Yet events exogenous to the bargaining process can break the resultant logjams, providing windows of opportunity for those concerned with the development of coherent and socially desirable regimes. This is particularly true when the exogenous forces come together to produce what can be portrayed credibly as a crisis. Consider, in this connection, the impact of the precipitous decline in fur-seal populations in the Bering Sea in the early twentieth century, the drastic reduction of blue-whale stocks in the 1930s, the discovery of an ozone hole over

---

45. For discussions of bargaining impediments see Schelling, *Strategy of Conflict*, and Oran R. Young, *The Politics of Force: Bargaining during International Crises* (Princeton: Princeton University Press, 1968).

Antarctica in the 1980s, or the nuclear accident at Chernobyl in 1986.[46] As these examples suggest, the resultant windows of opportunity may emerge quite suddenly, and they certainly do not last indefinitely. It follows that those interested in institutional design must be prepared to recognize occurrences of this sort quickly and to seize such opportunities promptly. Among other things, this means that there is a persuasive case for devoting time and energy to thinking systematically about the fundamental character and the relative merits of alternative regimes in specific issue areas, even when the existence of apparent stalemates suggests that the relevant negotiations are going nowhere. Only in this way can we be sure of having carefully thought through options available to push to the forefront when fleeting windows of opportunity appear.

Expand Scope

From the point of view of institutional design, there is much to be said for structuring negotiations over regimes in such a way as to provide the participants with what Brennan and Buchanan call "choices among rules of social order that are generally applicable and guaranteed to be quasi-permanent."[47] The resultant "veil of uncertainty" makes it difficult for individual parties to project the precise implications of specific arrangements for their own interests so that the "scope for potential agreement on rules is . . . wider than that for agreement on outcomes within specified rules."[48] Consider, by way of illustration, the Arctic shipping problem.[49] So long as the problem is cast as a narrow question of jurisdiction over the Northwest Passage, the United States and Canada will confront each other in a win/lose situation. Expanding the problem to encompass the articulation of a regime governing the activities of all vessels off the coast of Alaska and in Greenlandic waters as well as in the waters of the Canadian Arctic Archipelago, however, transforms this situation substantially. Faced with a range of concerns broadened in this manner, the parties must consider a larger set of interests and, therefore, weigh more complex motives. The same arrangements that would serve to secure their interests in the narrower

46. For relevant background see George L. Small, *The Blue Whale* (New York: Columbia University Press, 1971), and James G. Titus, ed., *Effects of Change in Stratospheric Ozone and Global Climate* (Washington, D.C.: U.S. Environmental Protection Agency, 1986).

47. Brennan and Buchanan, *Reason for Rules*, 30.

48. Ibid., 29.

49. Franklyn Griffiths, ed., *The Politics of the Northwest Passage* (Montreal: McGill-Queen's University Press, 1987), and chapter 7 above.

context may well prove unattractive with respect to the larger set of activities to be covered by the resultant regime. Under these circumstances, the parties will find it in their interest to adopt a broader perspective on the suitability of proposed institutional arrangements rather than simply focus on the pursuit of certain proximate interests.[50] As Brennan and Buchanan put it, "To the extent that a person faced with constitutional choice remains uncertain as to what his position will be under separate choice options, he will tend to agree on arrangements that might be called 'fair' in the sense that patterns of outcomes generated under such arrangements will be broadly acceptable, regardless of where the participant might be located in such outcomes."[51]

Multiply Roles

Similar comments are in order about Rawls's concept of a "veil of ignorance."[52] Those facing a "veil of uncertainty" know what roles they occupy in the relevant social systems. They simply find it hard to foresee the consequences of institutional arrangements because of the generality and longevity of the rights and rules included in various regimes. In Rawls's vision, by contrast, participants do not even know what roles they occupy. Strictly speaking, of course, any such situation must be purely hypothetical. Parties engaged in negotiations about the provisions of international regimes invariably have some sense of the roles they occupy. Yet there are circumstances in which the Rawlsian idea of a "veil of ignorance" has definite meaning under real-world conditions. These are situations involving several distinct roles, any of which individual parties may occupy under circumstances covered by institutional arrangements. Consider, for example, the nuclear-accident problem in which individual parties may occupy the roles of site, victim, or fortunate bystander in specific nuclear accidents.[53] Without doubt, the existence of a situation of this sort greatly facilitates efforts to negotiate coherent and socially desirable institutional arrangements. Faced with a situation where conditions approximating a "veil of ignorance" obtain, negotiators will find themselves with little to gain from narrowly self-interested initiatives and with clear-cut incentives to devise arrangements that serve to maximize the common good. To the extent that

50. See also Cynthia Lamson and David VanderZwaag, "Arctic Waters: Needs and Options for Canadian-American Cooperation," *Ocean Development and International Law,* 18 (1987), 49–99.
51. Brennan and Buchanan, *Reason for Rules,* 30.
52. Rawls, *A Theory of Justice,* 136–142.
53. For a fuller account see chapter 6 above.

there is room to structure the character of negotiations relating to international regimes, therefore, those interested in institutional design will find it worthwhile to make every effort to portray specific cases as situations in which the parties cannot be certain of occupying a single, well-defined role.

Emphasize Integrative Bargaining

By its nature, bargaining involves what Schelling has aptly character-ized as mixed motive interaction.[54] Those engaged in such interactions invariably seek outcomes lying as close as possible to their end of the contract curve. But they also have clear-cut incentives to avoid outcomes of no agreement and to expand the size of contract zones in such a way as to improve the payoffs for all parties concerned. Because it is easy to model behavior aimed at shifting the ultimate locus of agreement along a well-defined contract curve, there is an understandable tendency to focus on the resultant distributive bargaining in the analysis of mixed motive interactions.[55] But this tendency, which highlights the con-flictual aspects of bargaining, is particularly inappropriate in thinking about the formation of institutional arrangements. This is due partly to the fact that it is seldom, if ever, feasible at the outset to identify a well-defined contract zone in the complex bargaining involved in negotiating the provisions of international regimes. In part, the inappropriateness of focusing on distributive bargaining stems from the fact that those endeavoring to develop institutional arrangements typically see them-selves engaging in integrative bargaining as they hammer out agree-ments governing some class of interactions from which all parties are expected to benefit.[56] Nonetheless, it is easy to fall into the habit of bargaining in a distributive mode, deploying a range of tactical devices in the interests of driving a hard bargain, even in negotiations over the terms of regimes. No doubt, some of this behavior is inevitable in every negotiation dealing with the provisions of institutional arrangements. But there is much to be said for making a concerted effort to cast negotiations over international regimes as processes of integrative bar-gaining rather than distributive bargaining.[57]

54. Schelling, *Strategy of Conflict*.
55. See the various models surveyed in Young, *Bargaining*.
56. See also the perspective developed in Friedrich A. Hayek, *The Political Order of a Free People*, vol. 3 of *Law, Legislation, and Liberty* (Chicago: University of Chicago Press, 1979).
57. For another account that seeks to deemphasize distributive bargaining, described as positional bargaining, see Roger Fisher and William Ury, *Getting to Yes: Negotiating Agree-ment without Giving In* (Harmondsworth: Penguin, 1982).

Simplify Implementation

Regimes accepted on paper often end up as dead letters. This has been the fate of many of the regional-fisheries arrangements and some of the international commodity regimes.[58] To some extent, this fate can be attributed to the fact that international institutions exhibit the characteristics of public goods and therefore give rise to free-rider tendencies on the part of their members.[59] But there are several more specific factors that frequently play important roles in determining the degree to which institutional arrangements are implemented. Arrangements that are loosely textured and leave much to be worked out in administrative arenas, like some of the commodity regimes, are often crippled by various forms of bureaucratic politics, which arise regularly in implementation processes.[60] Regimes that require an elaborate administrative apparatus, like the proposed arrangements for the deep seabed, run the risk of being insufficiently supported and failing to get off the ground. Similarly, arrangements encompassing rules that make compliance difficult to verify, like some of the arms-control regimes, are easily undermined by reinforcing cycles of distrust on the part of those whose interests are at stake.[61] Understandably, more complex arrangements may seem appealing to those concerned with matters of equity or protecting certain interests (for example, preexisting producers in the case of deep-seabed mining). But the choice of complex arrangements to achieve such objectives generally leads to a widening gap between the ideal and the actual with respect to the operation in international institutions. Accordingly, those concerned with institutional design at the international level will find that there is much to be said for devising arrangements that can produce desired results and simplify processes of implementation at the same time.

Mobilize Leadership

Complex bargaining of the type under consideration here commonly places a premium on the exercise of leadership. In this connection, leadership is not so much a matter of some powerful actor structuring

58. J. A. Gulland, *The Management of Marine Fisheries* (Seattle: University of Washington Press, 1974), and Zacher, "Trade Gaps."

59. Olson, *Logic of Collective Action*. Public goods, in contrast to private goods, exhibit the characteristics of nonexcludability and jointness of supply.

60. Allison, *Essence of Decision*.

61. For a more general account of compliance in international society see Oran R. Young, *Compliance and Public Authority: A Theory with International Applications* (Baltimore: Johns Hopkins University Press, 1979).

the incentives of others through the use of negative and positive sanctions, though such efforts undoubtedly do occur on a regular basis,[62] as the deployment of entrepreneurial skills or creativity in identifying and putting together mutually beneficial deals. As the recent efforts of personnel from the International Atomic Energy Agency in connection with arrangements governing radioactive fallout and from the United Nations Environment Programme in connection with arrangements to protect the stratospheric ozone layer indicate, leadership of this sort need not revolve around the deployment of power in any conventional sense.[63] Characteristically, leaders in the negotiation of international regimes play important roles in seizing opportunities generated by exogenous events, structuring bargaining processes to focus on integrative rather than distributive issues, and putting together deals or packages of provisions that offer enough attractions to all parties to elicit their support. This, too, may seem quite removed from the image of rational planning that we often associate with institutional design or social engineering. Yet given the close relationship between institutional design and collective action, which I have described in the preceding section of this chapter, those concerned with the design of international regimes must focus their attention on efforts to exercise the sort of leadership under consideration here. As it happens, both an analysis of the nature of complex bargaining and an examination of actual negotiations regarding the provisions of regimes make it clear that this type of leadership holds considerable potential for those interested in the design of appropriate international regimes.

CONCLUSION

We might react to the preceding account of institutional design at the international level, with its emphasis on impediments or limitations as well as opportunities, as another affirmation of the significance of institutional arrangements as determinants of collective outcomes in international society. If prevailing arrangements could be altered more or less at will in the interests of achieving desired outcomes, after all, international regimes would emerge as epiphenomena; they would not

---

62. For some suggestive, though preliminary, comments on leadership in connection with international regime formation see Charles P. Kindleberger, "Dominance and Leadership in the International Economy," *International Studies Quarterly* 25 (1981), 242–254.

63. The efforts of Dr. Mustafa Tolba of the United Nations Environment Programme in facilitating the negotiations that led to the signing of agreements to control ozone depletion in 1985 and 1987 are particularly striking in this regard.

constitute important factors in explaining the nature of collective outcomes at the international level. For those committed to the argument
that institutional arrangements are important as independent variables,
therefore, much of the argument of this chapter may seem reassuring.

Yet there are persuasive reasons not to discard the idea of institutional
design with regard to international regimes. It is easy to imagine those
benefitting from the status quo, such as the current great powers or the
economic "haves," seizing upon the constraints discussed in the preceding sections to justify existing institutional arrangements and to dis-
courage the search for preferred alternatives. What is the point of
expending energy on the search for improved international regimes or
orders if efforts to bring about change are unlikely to succeed and are
apt to eventuate in unforeseen and unintended consequences even
when they do? It is a short step from this line of reasoning to the
conclusion that existing regimes are justifiable merely because they
seem to be securely in place. What is more, abandoning the idea of
institutional design with regard to international regimes would inevitably diminish the sense of efficacy of many who care about international
issues, thereby encouraging attitudes of fatalism concerning the future
of international society.[64] Whatever the intellectual bases for attitudes of
this sort, they are surely troubling from the perspective of all who are
genuinely concerned with the pursuit of the common good at the international level. A society whose members have a diminished sense of
efficacy regarding their ability to affect their collective fate is vulnerable
to the actions of a wide range of illiberal movements and poorly situated
to erect effective barriers against the occurrence of highly destructive
events, such as escalating violence or disruptive changes in the natural
environment.[65]

64. For a similar argument regarding domestic society see Brennan and Buchanan,
Reason for Rules, chap. 9.
65. On this theme, see also Leo Strauss, "What Is Political Philosophy? The Problem of
Political Philosophy," *Journal of Politics* 19 (1957), 343–355.

Index

Cornell Studies in Political Economy

EDITED BY PETER J. KATZENSTEIN

Library of Congress Cataloging-in-Publication Data

Young, Oran R.
  International cooperation: building regimes for natural resources and the environ-
  ment/Oran R. Young.
     p.   cm.—(Cornell studies in political economy)
  Includes index.
  ISBN 0–8014–2214–0 (alk. paper)
  ISBN 0–8014–9521–0 (pbk.: alk. paper)
     1. Natural resources—International cooperation. 2. Environmental policy—Interna-
  tional cooperation. I. Title. II. Series.
  HC59.Y685 1989   333.7—dc19   88–19078